American Institute of Chemical Engineers

22nd Annual International Conference of the Center for Chemical Process Safety

CCPS 2007

Held at the AIChE Spring Meeting 2007

April 22-27, 2007
Houston, Texas, USA

Printed from e-media with permission by:

Curran Associates, Inc.
57 Morehouse Lane
Red Hook, NY 12571
www.proceedings.com

ISBN: 978-1-60423-310-0

Copyright and Disclaimer

Process safety excellence is a vision that we all share and achieving this vision is challenging. There are many factors that impact process safety excellence. These range from sound process design, including inherent safety considerations, through reliability and inspection programs, adherence to good operating procedures, managing the risk of hazards to emergency response and community relations. The theme for this year's 22nd Annual International Center for Chemical Process Safety Conference is Advancing Process Safety through Design and Operations. I'm happy that many of these areas will be discussed, including an introduction of CCPS's approach to Risk Based Process Safety.

I want to thank you for your interest in this year's conference and hope that you will be able to employ the learnings from these presentations and discussions to advance towards the vision of process safety excellence within your organization.

Jim Muoio
2007 CCPS Conference Chair
Consulting Engineer
Process Risk Management
Lyondell Chemical
Houston, Texas
April 2007

22nd Annual International Conference of the
Center for Chemical Process Safety
2007

TABLE OF CONTENTS

16b Examining Organizational and Safety Culture Causes of the BP Texas City Refinery Explosion

Mark Kaszniak, U.S. Chemical Safety and Hazard Investigation Board, 2175 K St NW Suite 400, Washington, DC 20037

Abstract

On March 23, 2005, a series of explosions and fires occurred at the BP Texas City refinery during the startup of an isomerization (ISOM) process unit. Fifteen workers were killed and about 180 others were injured. However, the catastrophic incident on March 23rd was not the only major incident at the Texas City refinery. As part of its investigation, CSB learned that over the past 32 years, the BP Texas City refinery has had 39 fatalities. This discovery prompted CSB to investigate the organizational and safety culture precursors that produced the multiple safety system deficiencies that lead to the ISOM incident and the history of major accidents and fatalities at this refinery.

This paper will examine the organizational and safety culture precursors rooted in the refinery's history. The paper will also show how these organizational issues extended beyond the ISOM unit and the Texas City refinery to the corporate oversight system of BP Global and to the 1999 merger between BP and Amoco Corporation, where the safety impacts of major organizational change were not effectively reviewed.

Disclaimer

This paper has been prepared for general informational purposes only. This paper represents the individual views of the authors and all references, conclusions or other statements regarding current on-going CSB investigations are preliminary in nature and limited to information that is already in the public domain. Furthermore, this paper is not a product of the Board and its contents have not been reviewed, endorsed, or approved as an official CSB document. For specific and accurate information on completed investigations, please refer to the final published investigation report by going to the CSB website at www.csb.gov and clicking on the specific report desired under "Completed Investigations." To the extent this paper includes statements about the conclusions, findings, or recommendations of the Board, such statements come under the general prohibition in 42 U.S.C. µ7412(r)(6)(G).

FINDING POTENTIAL FAILURES DELIBERATELY

Jack Hipple
Innovation-TRIZ
Tampa, FL
813-994-9999
www.innovation-triz.com
jwhinovator@earthlink.net

Our concerns for preventing and eliminating the causes of chemical and safety incidents that affect human life and cause property damages have led us to use comprehensive checklists, such as HAZOP. This type of checklist exhaustively analyzes a proposed or operational process based upon its design conditions and questions each of its design or operational variables and asks the basic question in different ways, "What if the design point is not where it is supposed to be"? Does it matter? If so, to what extent? What is the importance of the degree of variability from the design point? These considerations are especially important during startups and shutdowns when normal operational conditions are seldom present. Despite these checklist efforts, fire, explosions, and loss of life and property still occur in the chemical industry. Most often, these accidents are the result of violations of defined procedures. In many cases, however, the cause is found to be due to a condition or procedure that was not reviewed or discussed previously. This learning is then incorporated into the process knowledge and the identical incident may not occur again. This paper reviews a unique process for analyzing processes and operational procedures for potential failure mechanisms that is capable of identifying potential failure routes and mechanisms in a more comprehensive fashion, using a reverse saboteurial thinking process based upon a unique Russian inventive problem solving process known as "TRIZ" (Theory of Inventive Problem Solving).

1. TRIZ Background

1.1 History

In the early 1950's a brilliant Russian patent examiner developed a powerful and unique problem solving process and algorithm for solving difficult engineering design problems that had, at their root, what appeared to be irresolvable contradictions. In other words, performance or design parameters that were in direct conflict. An example in today's world would be an automobile that is both heavy (for comfort and injury protection) and light (for fuel economy). Genrich Altshuller, himself a prolific young inventor, worked for the Russian navy and examined patent disclosures in the same sense as a current day patent examiner. In reviewing hundreds of thousands of patents, he sought out the top 2-3% that were truly breakthrough and novel, and analyzed these few to determine the basic nature of the inventive principle that was used in the invention, regardless of the field of science or industry from which it came. His thought process for determining which patent to put in this category was not only that the invention was truly unobvious, but the fact that invention resolved a major design or operational contradiction.

In this process, he discovered an amazing thing--there were a limited number (40 to be exact) inventive principles that were constantly re-used across virtually all engineering and technology areas. His work and that of his students to follow resulted in problem solving tables, problem solving tools, and algorithms that are still in use today and being adopted by more and more companies, including Motorola, Dow Chemical, Shell, Unilever, Boeing, S.C. Johnson, and General Electric. In many cases, TRIZ is incorporated within the Six Sigma and Design for Six Sigma processes and programs to provide the problem solving tools that these outstanding problem definition tools lack. It is also slowly being incorporated into graduate engineering curricula in engineering schools in the US. It has become a mandatory part of the entire educational system in France. Continued study of the global patent literature has failed to identify additional inventive principles and has extended its application to fields such as biological systems.

The tool kit and its components came to be known as "TRIZ" (Theoria Resheneyva Isobretatelskehuh Zadach), or "Theory of Solving Inventive Problems", occasionally also translated as "Theory of Solving Problems Inventively". Within the tool kit are problem solving tools such as the Contradiction Table, 40 Inventive Principles, Separation Principles, Lines and Patterns of Evolution, and ARIZ ("Algorithm for Solving Inventive Problems").These tools can be used separately or within the context of the overall algorithm.

1.2 Algorithm and Problem Solving Model

Over time, the TRIZ process has evolved into a problem solving algorithm, commonly known as ARIZ (algorithm for solving inventive problems). This algorithm has had refinement over time and is described as ARIZ-85, ARIZ-91, etc. to reflect the year in which it was finalized. These changes have been subtle over time. The simplest version of this algorithm is as follows:

1. Decide on the problem focus area (a key aspect)
2. State the ideal result for the resolution of the problem
3. Evaluate and identify resources within the system, including defects and deficiencies that may allow achievement of this ideal state. Resources from a TRIZ perspective is a very broad category and includes not only materials and the surrounding environment, but fields and converted fields (electrical produces a magnetic field whether we like it or not!), time (before, during, and after), space (outside, inside, voids), functional and field conversion (abrasion always generates heat)
4. Identify contradictions standing in the way of achieving the ideal state
5. Resolve these contradictions through the use of TRIZ principles.

When applying this algorithm, it is important to look not only at the system of concern, but also the sub-systems and super-systems that it incorporates and is a part of.

Let's take a simple example to illustrate this thinking and analytical process. A manufacturer of expensive bathing soap uses an assembly line for packaging the soap into boxes. Despite all the best efforts of the engineers who designed this process, an empty soap box occasionally passed completely through the process and appeared on the shelf at the grocery store, resulting in returns and customer complaints. The engineering staff decided to solve this problem by installing a "see through" X-Ray system that would detect an empty box and then remove it with a robotic based mechanical system. This system no doubt would have worked, but obviously would have been involved major capital investment and maintenance costs. While the engineers were designing this complicated system, the night shift operator found an existing, high flow floor fan and moved it next to the process line. Any empty soap box simply blew away. Though this night shift operator did not have TRIZ training, he indirectly used the algorithm:

1. Problem focus area: resolve the problem on the factory floor. This is an important point and does not lend itself to right or wrong discussions. This problem conceivably could have been addressed and solved at the point of insertion of the soap bar into the box, or downstream at the grocery store or at the warehouse.

2. The ideal state: the empty soap boxes separate themselves out. Again, this ideality or ideal final result (IFR) statement would have been different if another focus area had been chosen. It is worth noting that this type of ideality statement is extremely difficult for experienced adult engineers to say and express, as mentally this seems to be impossible to do at first glance so the goal is narrowed to something like "80% of the defective boxes are removed with the addition of a new system".

3. Identify resources: the components of the assembly line, the box, the operating personnel, utilities, etc. But most important, we need to list the empty box itself as a resource even though it is actually the problem. The empty box is characterized by void, lack of weight, etc. It is important to state these resources in physical parameters, not with jargon that may be unique to the process or product.

4. Identify contradictions: the box should be heavy and contain soap and transparent, light, etc. to ease in its inspection and removal

This is far as we need to go in this case. The fact that the empty box is virtually weightless allows us to use this negative aspect of the system as a positive resource for the problem resolution.

2. Inverting the TRIZ Process and Algorithm: Predictive Failure Analysis™

In the late 1960's some of Altshuller's students came up with a very clever adoption of the TRIZ algorithm by "inverting" it. Instead of trying to solve problems that existed, the inverted algorithm could be used to identify potential problems before they became problems. Let's take the example[7] of the explosion of a kerosene storage tank containing a water layer and the concern for explosions within the tank. Our normal approach would be to do a HAZOP review of this system, making sure that no air was accidentally introduced into the tank, the tank was not overfilled, and pumps, alarms, and level controls systems were functioning and backed up. Despite having done this, the tank exploded. How could this have happened? The algorithm is as follows:

1. State the ideal result that is desired: the tank is safely operated with no spills or injuries
2. Invert the ideal result: the tank and storage system do not function properly (they leak once in a while and possibly cause a fire)
3. Exaggerate the inverted ideal situation: the tank never functions as desired. It explodes every time we use it, causing environmental damage and loss of life.
4. How would we accomplish this? What resources do we need? We are saboteurs and our goal is to make sure this happens!

In thinking about this, we obviously need a flammable material that will ignite at the ambient storage conditions. In this case, the river water, used as the source of the water layer, contained bacteria (levels of which might change with season of the year and discharges to the river upstream) which generated methane within the tank. It is doubtful that even a thorough HAZOP review would have picked this up as a concern area. With this process, we ask, "How would we make sure that the tank could explode all the time? What resources are necessary? How would we make sure they entered the system—all the time, every time?"

In addition to unobvious material resources, time is frequently a resource for disaster that is ignored. Tank welding and maintenance are classic examples where "reverse" thinking can prevent accidents. An example[7] is an acid tank drained for welding work that was delayed for over 40 days. Residual acid attacked the metal and generated hydrogen. Obviously, the tank should have been rechecked, but we can also look at this Predictive Failure Analysis™ algorithm:

1. Ideal Result: the tank is weldable at any time
2. Inverted Ideality: the tank occasionally contains flammables and must be monitored and checked
3. Exaggerated inverted ideality: the tank ALWAYS explodes when welded
4. What resources are required? (flammable gas and oxygen)

How would we generate a flammable gas within this system? Residual acid over time (a resource!) attacks the metal, generating hydrogen gas.

There are numerous other examples that have been documented, where this inverted algorithm could prevent major accidents:

1. Work permits not in process areas. How would we ***make sure*** that a maintenance team opened a manhole cover and were exposed to process fluids?
2. Mislabeling of tanks, materials, cylinders. How would we ***make sure*** that the wrong materials were used? This is more complete and drastic than simply reviewing the existing or planned labeling process or protocols.

In our experience with this process, a group never fails to identify additional routes to failure than were identified with checklists. Especially significant is the group's ability to identify previously not thought of resources that could contribute to failure.

It is worth noting that this mental process could have been applied to the empty soap box problem previously discussed. How would I *make sure* that a soap package *never* contained any soap?

3. Case Study Examples

3.1 Salt Evaporator Fatality

A large salt solution evaporator was shut down for welding of several leaking tubes. Normal operating conditions were approximately 140°C. The proper procedural documents were completed and the vessel inspected for air presence prior to entry of a welder. However, it was necessary to allow the evaporator to cool prior to human entry. When the welder entered the tank many hours later, he collapsed almost immediately and died from lack of oxygen, despite the previous air monitoring test. No one could reach him in time without further injury resulting.

If we look at the Predictive Failure Analysis™ algorithm for this problem, it looks like this:

1. Define the ideal state: the welding process goes perfectly with no accidents or injuries
2. Invert the ideal state: The welding process fails and injuries occur (there may several different ways of expressing this)
3. Exaggerate the ideal state: ***All*** entries to the vessel injure or kill personnel
4. What resources are required to accomplish this?

Now of course, the evaporator tank should have been re-sampled upon entry due to passage of time and that could have prevented the fatality, but is it probable that those doing the re-inspection would have given serious thought as to why the oxygen level dropped? Why it could drop? In order to remove oxygen from the tank, we need a chemical reaction of some sort. In this case, the elevated evaporator temperature allowed the iron tubes to react with the oxygen in the air, producing iron oxide and depleting the tank of oxygen. We frequently forget that corrosion is a chemical reaction and has all the attributes (stochiometry, thermodynamics, energy release) of any other chemical reaction.

3.2 Sulfur Trioxide Release

As seen in Figure 1, a bulk SO_3 tank is being used to feed gas to a sulfonating process. Unloading into the tank is via T/T pumping and the level in the tank is monitored by a simple bubbling device. The air used in this bubbler is scrubbed through a sulfuric acid scrubber, controlled in the 93-98% range to prevent solidification as well as optimum scrubbing. The scrubber vents to the atmosphere. In actual fact, this scrubber's acid was allowed to drift out of specification on the high side, solidifying the acid and allowing gas to escape directly into the atmosphere, causing a community alert and local evacuation.. This process had been HAZOP'ed several times within a two year period.

This failure analysis process was applied in this situation to assess its effectiveness in potentially improving process safety reviews. The algorithm was applied as follows:

1. The ideal state: Liquid is contained within the tank at all times and no releases, injuries, or environmental damage results from the presence of the tank
2. Inverted ideality: Liquid escapes from the tank
3. Exaggerated inverted ideal state: SO_3 leaks from the tank at all times and causes injury, environmental consequences, withdrawal of plant operating permit, and jail terms for the site management.
4. What resources do we need to ensure this happens all the time?

When this question was first asked of the group, 3-4 people made comments such as "high pressure", "high temperature", or "relief valve opens". Finally a participant said "a hole". This was a very profound recognition of the necessary root cause. Without a hole, a leak cannot happen, no matter what other conditions prevail. When we look at the flow sheet, the hole at the top of the scrubber is obvious, so we ask "why is the hole there?" The answer to that is to provide venting for the nitrogen used as a bubbling gas differential pressure measurement. Is that the only way to measure level in this tank? The answer to that is "no", as the group recognized. With SO_3 being a paramagnetic fluid, an externally mounted nuclear magnetic resonance level system is usable, eliminating the need for the continuous vent. A HAZOP review on this process analyzes the process as designed and operated and does not necessarily question the fundamentals of the design.

Figure 1: SO_3 Handling System

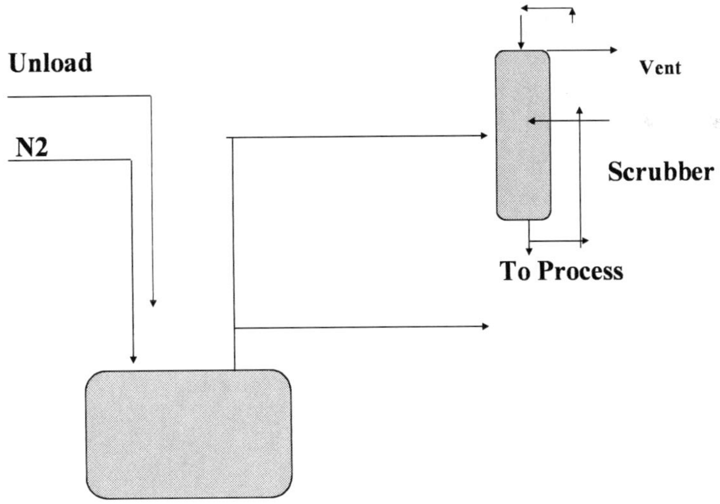

3.3 The Leaking Bromine Drum

The United States, not being a member of the common market, has difficulty economically shipping some extremely corrosive materials, such as liquid bromine, that normally require expensive nickel based alloy metals and totally dry conditions. A major chemical company decided to overcome this obstacle by designing an extremely thick steel container that would survive the time required for overseas ship transport and then distill the bromine away from its $FeBr_3$ corrosion products in a simple one stage distillation, upon arrival in Europe, prior to final packaging in the normal corrosioin resistant drums. This process was implemented and worked flawlessly the first time, following all the normal "check listing" regarding dryness, drum fill %, etc.

During the second shipment, however, a number of drums ruptured in transit, causing a significant incident aboard an ocean transport ship. When the facts of this incident were reviewed, it was determined that some of the drums had in fact been overfilled to near 100% and that some of the product contained more water than specification. Extra time in a southern US port in the summer was also a contributing concern. These facts alone, however, were insufficient to explain the corrosion and/or rupture of the drums, based on their pressure rating and construction.

How could we make *sure* that these drums ruptured *every* time they were shipped? What resources are required? Excessive pressure is certainly one such resource and the summer temperature and long port hold times involved in this shipment contributed to this possibility. However additional heat and pressure was needed. Where could it come from? As mentioned earlier, bromine is very aggressive to steel, but in a steady, non-catastrophic way. We occasionally forget that corrosion is chemical reaction and has heat effects in the same way any other chemical reaction does. Normally, the magnitude of this heat is extremely minor relative to other heat effects in a system..

Let's take a look at a basic overview of what was going on inside some of these drums. We have a small scale chemical reaction going on. Iron is being converted to iron bromide and the heat generated from this process is released through the drum walls in the same way it would be in an intentional reactor. If this heat generation is greater than the heat removal rate, the liquid contents will rise in temperature, eventually causing vaporization and a significant pressure rise in a vessel not designed to contain high pressures. Qualitatively, this looks like figure 2:

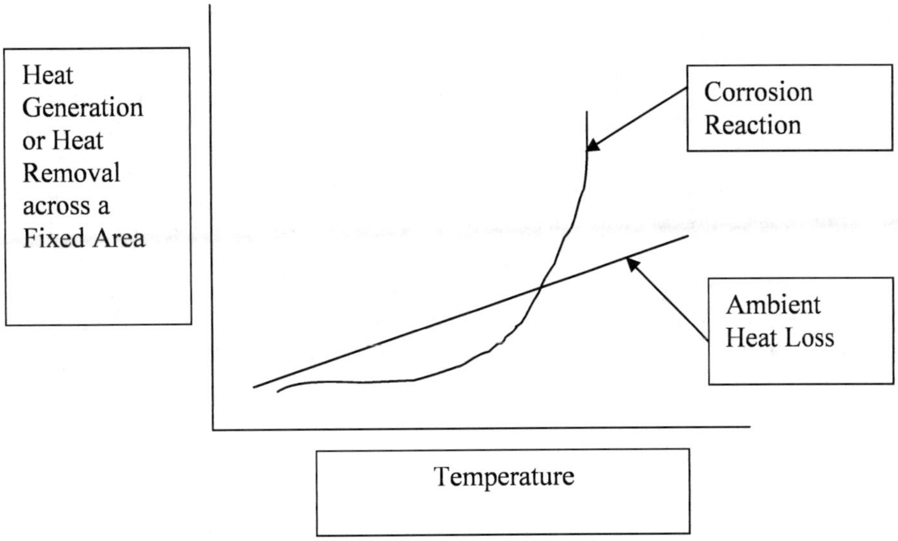

Figure 2
Kinetics vs. Heat Transfer

The other factors, in addition to bromine's boiling point of 60°C, is the fact that bromine's heat capacity is $1/10^{th}$ that of water or approximately 0.1 cal/gm.-°C vs. approximately 1.0 cal/gm.-°C for water. When this system was re-analyzed incorporating these additional factors mentioned previously, the day of release on board the ship was nearly exactly calculatable.

How does this algorithm apply?

1. Ideal State: the drums arrive overseas without leakage or incident
2. Inverted ideal result: the drums leak occasionally, requiring special precautions or procedures.
3. Exaggerated inverted ideal result: the drums leak on ***every*** shipment causing injury and loss of business opportunity
4. How do we make this happen? What resources might be available? In this case we have material contamination, pressure, low heat capacity, heat generation from the corrosion reaction, and insufficient heat removal

4. Potential Applications

Even though we have focused on some chemical examples, this process has been used to identify previously unseen causes in other areas. For example:

1. Software code development: How do we make sure that the software ***never*** functions?

2. Food processing and contamination: How do make sure that ***every*** food batch is contaminated and causes injury or death in the retail grocery and restaurant chains?

3. Management processes: How do make sure that management communication is ***never*** heard by employees? How do make sure that ***no one*** wants to stay employed with this company?

We believe that this Pro-active Failure Analysis™ process can bring great value to hazardous chemical processing analysis, in addition to contributing to error reduction in other aspects of a commercial enterprise.

References

1. Altshuller, G.S.: And Suddenly the Inventor Appeared, Technical Innovation Center, 1996
2. Altshuller, G.S.: TRIZ Keys to Technical Innovation, Technical Innovation Center, 1997
3. Hipple, Jack: Analyzing Failures Using TRIZ: Part 2 of 3 part series in Chemical Engineering Progress, May 2005
4. Hipple, Jack. TRIZ: Inventive Problem Solving, ASME/AIChE training class, www.asme.org
5. Hipple, Jack: How TRIZ Will Affect the Future of Problem Solving, Futures Research Quarterly, 2003
6. Kaplan, Stan, et.al. New Tools for Failure and Risk Analysis, Ideation International, 1999
7. Kletx, Trevor. What Went Wrong: Case Histories of Process Plant Disasters, 4th ed. Gulf Publishing, 1998
8. Mann, Darrell: Hands on Systematic Innovation, Creax Press, 2002

Websites:

www.aitriz.org
www.innovation-triz.com
www.triz-journal.com
www.etria.net

™Predictive Failure Analysis and PFA are registered trademarks of Innovation-TRIZ

Quantification of Human Factors for Quantitative Risk Analysis

Dr.-Ing. Katharina Löwe and Dipl.-Ing. Gitahi Kariuki

Technische Universität Berlin, Institute of Process and Plant Technology,
Berlin, Germany
email: katharina.loewe@tu-berlin.de

ABSTRACT

Inclusion of human factors (HF) in quantitative risk analysis (QRA) is done by use of human reliability analysis (HRA) techniques. However, HRA methods experience some drawbacks, which lead to uncertainties in human error probabilities (HEP). These limitations lead to uncertainty in QRA results.

This work was founded on premise that most HEPs are inaccurate and therefore do not represent the actual conditions of the plant being analyzed. It is critical that potential causes of human errors that may eventually lead to major accidents be exhaustively identified and quantified for a complete QRA. Unfortunately, the tools currently used by analysts for hazards identification still face some limitations. A systematic method to analyze the underlying human factors, which cause errors that lead to accidents, is needed. This work brings in the aspects of HF into QRA. It applies risk analysis framework and is extended to capture human and organization factors that influence the operator performance in order to identify the actual error producing conditions.

1. Introduction

In the recent past HF is a field that has attracted much attention as a way of reducing major accidents in many industries including nuclear and aviation. Chemical and petrochemical industries have joined this cue. The main reason being that many major accidents have been attributed to human error as a primary cause. Statistics show that in the chemical process industries accidents attributed, at least in part, to human error range between 60–90% [1, 2] and in the petrochemical industry e.g. oil refineries where automation is very high, human error accounts up to 50% [3]. Actually it is evident that the relative number of accident events due to human errors is on the rise while those due to technical failures are decreasing.

Many designs put a lot of emphases on engineering reliability. However, safety of a process plant is influenced by the quality of design, operational and organizational factors. To improve safety and therefore reduce undesired events requires designing of equipment, operations, procedures and work environments in such a way that they are compatible with the physical and cognitive capabilities and limitations of human beings. It involves identifying unrealistic demands on operators and maintainers by system characteristics [4]. Systems that do not adhere to this are doomed to failure because of unjustifiable demands to the facilities and operators that easily lead to unsafe situations.

Many approaches on human error focus on symptoms of human error rather than the underlying causes [5]. This is not only wrong and naive but also overly simplistic. It is like identifying symptoms of a disease without examining the underlying cause or further defining the disease. Unless we understand all these indirect factors that lead to direct human error there are slim prospects of reducing accidents or incidents caused by operator errors. This calls for a more systematic and comprehensive method for identifying the causal sequence of a human error event that would enable development of sound intervention strategies.

The objectives of this work as a whole are:

i) To develop a human factors classification and evaluation model.
ii) To develop a framework to identify human error events and to analyze the human and organizational factors behind these error events. This information is critical in establishing the influence that impacts indirectly yet powerfully on the probability of an accident.
iii) To quantify the human and organizational factors for quantitative risk analysis (QRA).

These steps are described in more details in the following sections.

2. Human Factors Approach

Most important is a framework that is able to unearth most if not all underlying causes of human error. The model that will be used is a detailed accident causation model. It explains how an accident propagates and the factors that directly and indirectly influence it.

2.1 Human Factors Definition and Background

As indicated in the introduction, statistics show that human failure is a major cause of undesired events in process industries. However, the role insufficient design plays towards these incident causations and the contributions the management failures have towards human error occurrence are not given the weight they deserve.

To sufficiently address human failure a human factors approach is necessary. Human factors refer to environmental, organizational and job factors, and human and individual characteristics, which influence behavior at work in a way that can affect health and safety [3]. It takes human as an integral part of plant design and procurement from the earliest stages. Figure 1 [4] shows that an undesired event is a result of latent conditions and active errors.

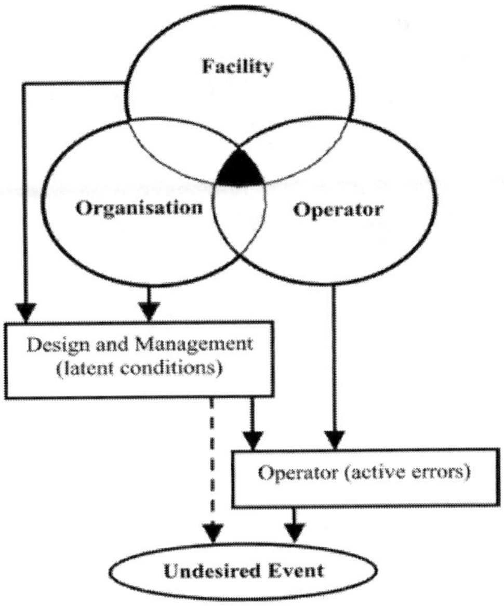

Figure 1: Human Factors contribution to undesired events [4]

2.2 Integrated Human Factors Model

Figure 2 shows the integrated human factors model which was developed for this work. An undesired event is caused by equipment failure, human error, external impact (not included in the figure) and a combination thereof. These are referred to as direct causes. The science of analyzing equipment reliability and failures has developed over the years. Many methods are in existence but none will be discussed here because they lie outside the scope of this work.

Inadequate design implies that the physical and cognitive capabilities and limitations of populations of people are not incorporated into design and operation of a system, process, or equipment. In this case technical design refers to human factors engineering/ human engineering. This branch of engineering is concerned with designing of products, processes and equipment used in manufacturing so as to maximize their ability to be used comfortably, safely and effectively by human beings [7]. Areas that are considered are workplace layout, workplace accessibility, controls and displays, workplace environment and labeling and signage. These areas are addressed using the human factors engineering approach described earlier. Management faults include inadequate training, procedures & procedure development, work schedules, staffing, shifts & overtime among others. From Figure 2 it can be seen that management faults influence equipment failure, human factors engineering and operator characteristics to some extent. An organization, through formulation of policies and safety culture has also a direct influence on the way a company manages its safety systems. This also affects the performance of the operator.

Organization, management systems and facility design, when inadequate, are the causes of latent conditions [6]. Latent conditions do not immediately affect the functioning of the system but in combination with other factors like active operator error and/or a local trigger (high temperature, high pressure) they could result to a disaster. On the other hand active errors manifest themselves immediately when they are committed.

If not recovered they could lead to undesired events (accidents or near-miss). Active errors are directly contributed to the operator and could be analyzed using the cognitive approach. Equipment failures can be analyzed through plant condition or maintenance and inspection management.

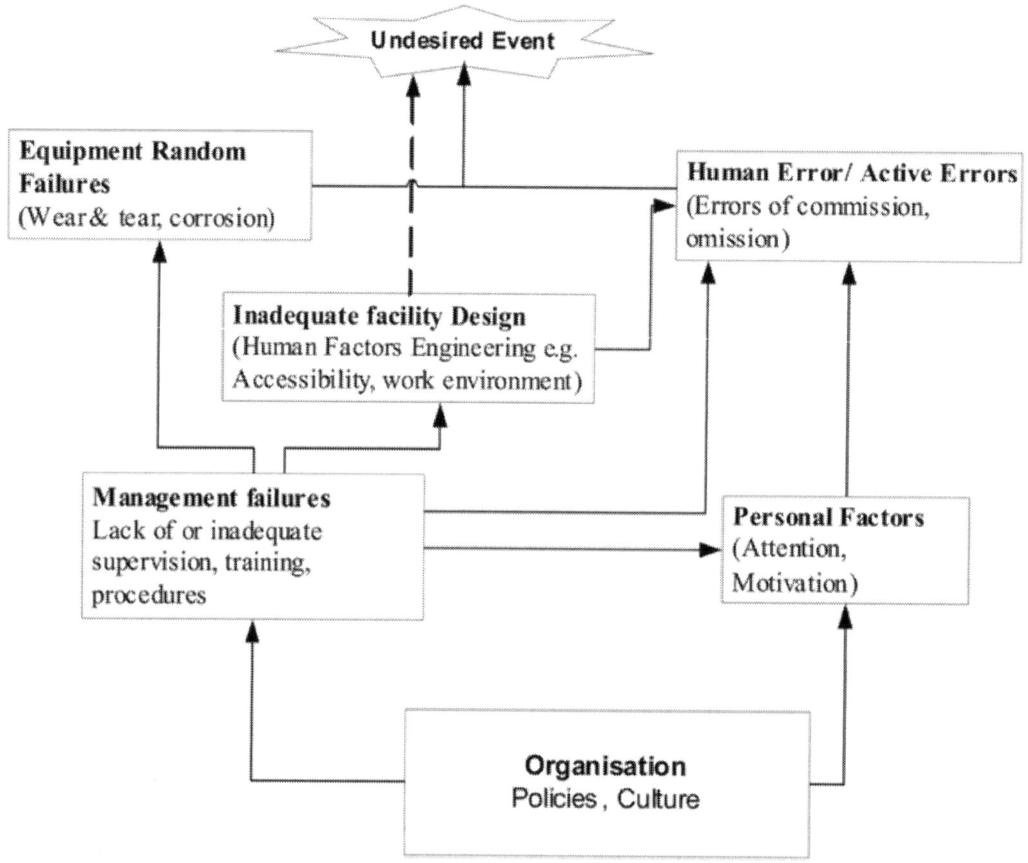

Figure 2: Integrated Human Factors Model [8]

3. Classification of Human Factors

Human and organizational factors that are known to have influence on the performance of the operator are put in various groups. One of the most challenging tasks was to reach a consensus on the most practical classification of the whole human factors domain. The whole spectrum is large and has been approached in different ways by different authors [9, 10]. Moreover, factors that may apply in other industries e.g. nuclear, medical and/ or aviation may not be applicable in the chemical process industry. The classification of human factors was achieved after a wide consultation with experts from the petrochemical and chemical process industry. This was initially done during the development of PRISM guidelines [11] and is illustrated in table 1.

Table 1: Human and organizational factors in a process plant [8]

Factors	Attributes
Organization (ORG):	A1 Human factors and safety policy A2 Organizational culture A3 Management of change A4 Organizational learning (audit & reviews) A5 Line management & supervision
Information (INF):	B1 Training B2 Procedures & procedure development B3 Communication B4 Labels & signs B5 Documentation
Job Design (JD):	C1 Staffing, work schedules C2 Shifts & overtime C3 Manual handling
Human System Interface (HSI)	D1 Design of controls D2 Displays D3 Field control panels D4 Tools (hand) D5 Equipment & valves
Task Environment (TE):	E1 Lighting /Illumination E2 Temperatures E3 Noise E4 Vibration E5 Toxicity
Workplace Design (WD):	F1 Facility layout F2 Workstation configuration F3 Control room F4 Accessibility
Operator Characteristics (OP):	G1 Attention/ motivation G2 Fitness for duty G3 Skills and knowledge

This classification has been validated twice. The first validation took place at Chinoin Co. Ltd in Hungary, member of Sanofi-Aventis Group, as a part of PRISM guidelines validation [10].

Furthermore for an easy and systematic human factors analysis of an industrial plant an assessment tool based on this classification was developed. The output of the technique is a qualitative and quantitative value for the overall quality of the human factors of the plant, as well as a detailed report of the state of each human factor and what could be done to improve or maintain their actual status [14].

The second phase of the validation took place as part of the practical evaluation of this software to assess quality of human factors. This was done at Bayer CropScience in their Frankfurt site in Germany. In this exercise it was agreed that the classification covered most of the factors that could affect human operator performance. But it was found that some attributes like "HF policy" need to be elaborated. In summary the factors outlined in table 1 are thought to have the biggest influence on the operator performance. If any of them is inadequate then there is a higher probability of human error event.

4. Background to the Quantification of Human Factors

This section will use a fault tree [12] in figure 3 to explain the necessary chapters. The purpose of human factors analysis is to assign specific failure probabilities for each initiating event based on different underlying human factors on a selected incident. The existing models of present QRA have been taken as the basis for this new methodology.

Step 1: Identifying the causes of an undesired event.

All the initiating events and basic events are identified using the conventional QRA methodologies. For this study a fault tree analysis (FTA) is used, see figure 3.

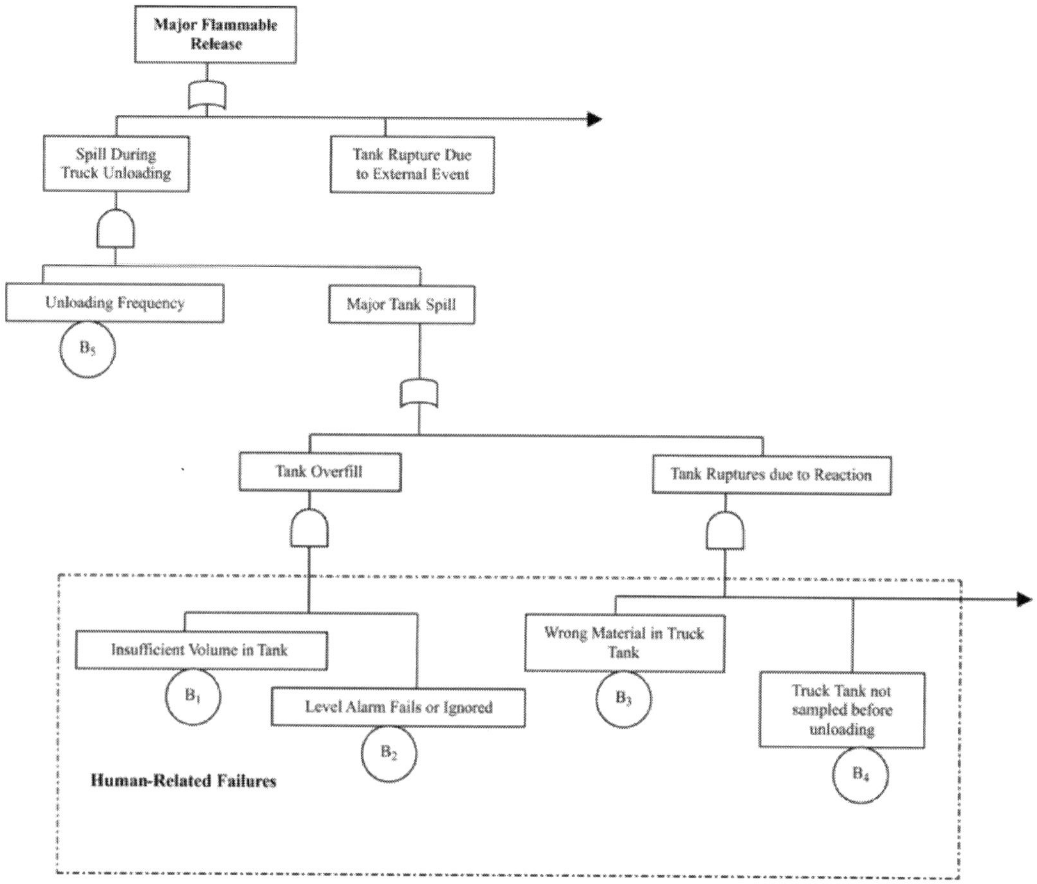

Figure 3: Fault tree Analysis of a Major Spill

The top event "Major Flammable Release" which has a frequency of 3.2×10^{-2}/yr could be caused either by:

Spill during truck unloading	3.0×10^{-2}/yr
Tank rupture due to external event	3.0×10^{-5}/yr
Tank drain ruptures	1.0×10^{-4}/yr
Tank rupture due to implosion	2.0×10^{-3}/yr
Tank rupture due to overpressure	2.0×10^{-5}/yr

This classification is irrespective of the type of failure. If we look closely it is clear that the event contributing most to the top event is spill during truck unloading (93.75%).

This is one of the reasons we will concentrate on analysis of the event "Spill during truck unloading". The other reason is that this event has human error events in it. This value (3.2×10^{-2}/yr) is the summation of all minimum cut-sets. Basic events B_1 (Insufficient volume in tank), B_2 (Level alarm fails or ignored) and B_5 (Unloading frequency) make one minimum cut-set which has a frequency value of $3 \times 10^{-2}\ yr^{-1}$. Minimum cut-sets are defined as sets of basic events that contain no redundant elements [12].

Step 2: Identifying initiating events with human error elements

This approach aims at coming up with a systematic methodology to analyze the situation that may lead to human error. The approach should be able to predict the conditions that support the occurrence of error. This means we are shifting our attention from the error itself but are focusing on the factors that support the occurrence of the error. From the fault tree analysis of flammable liquid storage tank, it was found that the following basic events contained human failure elements in them: Basic event B_1: Insufficient Volume in Tank; B_2: Level Alarm fails or ignored; B_3: Wrong Material fed into Tank; B_4: Truck Tank not sampled before unloading and B_5: Unloading Frequency.

These are typical events found during FTA. Not all basic events are classified as initiating events. Event B_1 may not be classified as an initiating event but for the top event to occur this condition (filling the tank) has to be present. Any time the condition is present then an opportunity for error occurs. It is referred to as enabling event [12].

Step 3: Identifying underlying human factor causes

A procedure to identify the underlying human factors is described in details in [8]. After the initiating events or hazardous conditions have been identified the human factors influencing these human error events are identified. One initiating event or basic event is analyzed at a time. Each factor is considered to determine how much it could influence the initiating event.

Table 2 illustrates a summary of factors that may have significant impact on basic events B1, B2 and B5. These have been identified using the procedure illustrated in [8]. It is worth to note that this exercise is subjective and therefore may require input from several experts to make it valid.

Table 2: Attributes having significant impact on basic events B_1, B_2 and B_5.

	Information	Supervision	Human-System Interface Design	Task Environment	Workplace Design	Operator Characteristics
Basic Events						
Insufficient Volume in tank	Procedures Documentation Training	Checks	Displays			
High level alarm fails or ignored	Training		Displays	Lighting	Accessibility	Attention / Motivation Skills and Knowledge
Unloading Frequency	Documentation Labels & signs	Checks				Attention/Motivation Fitness for duty

As an example, ordering procedures influence the volume that is available in the receiving tank. If documentation were not done adequately then ordering twice would be a feasible error. Job allocation determines the amount of workload on the operator.

5. Quantification of Human Factors and Integration into QRA

5.1 Analytic Hierarchy Process

The contribution of each human factor towards the occurrence of human error is calculated using Saaty's Analytic Hierarchy Process (AHP). This is a mathematical technique for multi-criteria decision making [8, 13]. It involves building a hierarchy (see figure 4) or ranking of decision elements and then making comparisons between each possible pair in each group as a matrix.

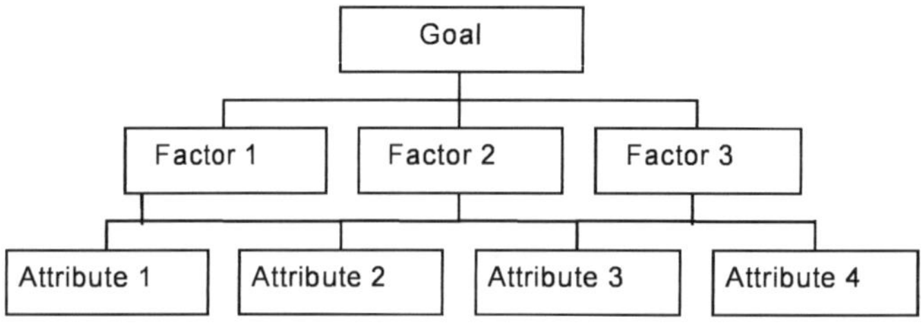

Figure 4: Generic hierarchical decomposition

This gives a weighting ω_i', for each element within a level of the hierarchy and also a consistency ratio, which is useful for checking the consistency of the data. A weight ω_i' is a fractional value between 0 and 1. The inputs of the matrix were obtained from experts within the European chemical process industry [14]. A questionnaire was sent out to 24 judges from the industry and this made the weighting results to be as practical as possible. The weights of attributes belonging to the same factor on same hierarchy level must sum to 1. AHP is especially suitable for complex decisions, which involve the comparison of decision elements, which are difficult to quantify [15].

The next step is to develop a rating system. The tool to be used by the assessor has to valuate various facility attributes. The rating value obtained is to be multiplied by the respective global weight ω_i' of this attribute. The separate contributions of each attribute to the overall objective (score) are considered to be additive. The human factors modification index β is defined as the sum of products of all individual attribute's weight and their respective performance measures [15].

$$\beta = \sum_{i=1}^{n} \omega_i' v_i(x_i) \qquad = \sum_{i=1}^{n} \omega_i' r_i' \qquad (1)$$

$v_i(x_i)$ is a value function of attribute x_i represented by a rating (referred to as performance measure) r_i'. The modification index β is a cardinal numerical score. The score trades off different levels of performance among attributes in a compensatory way, by using the cardinal weight ω_i' calculated using matrices of pair-wise comparison and by characterizing each attribute numerically.

5.2 Integrating HF into QRA

Once a potential human error event has been identified, the factors that could influence its occurrence are identified and quantified. Each factor will be assigned a weight ω_i', which represents how much it contributes towards human error occurrence. The weights help to calculate the human factors modification index β which is used to adjust the nominal human error probability to reflect the plant conditions. The procedure can be summarized as shown in figure 5.

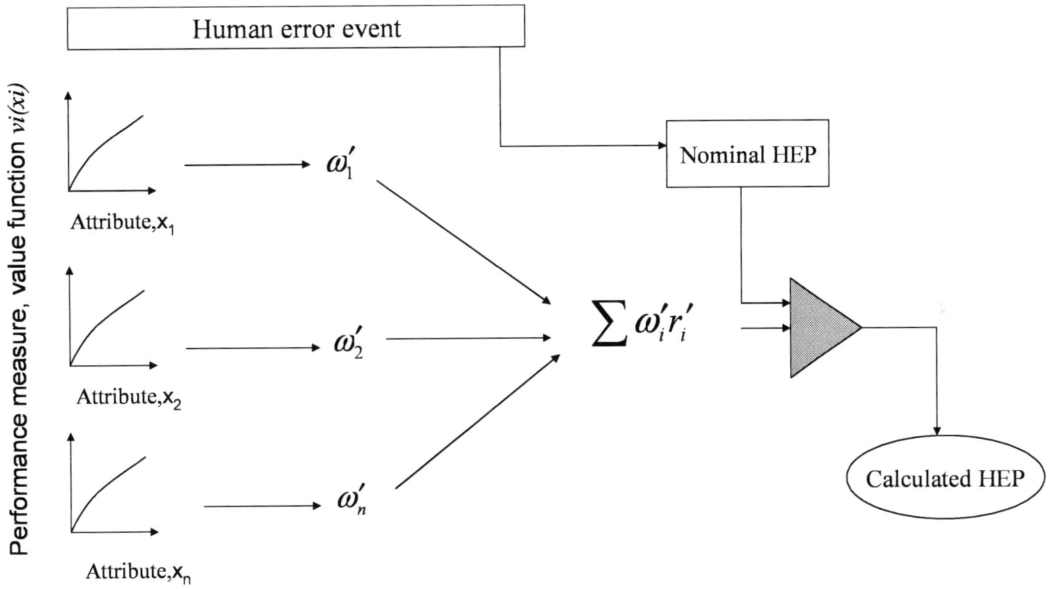

Figure 5: Modification of Human Error Probabilities by use of weighted performance influencing factors.

From weights ω_i' and performance measures r_i', the human factors modification index β for each human error event is calculated. The factors affecting initiating or basic events are identified before the quantification process starts. The factors influencing human performance have been identified as organization, information, job design, human machine interface, task environment, workplace design and operator characteristics and these are illustrated on table 1 and each of these has accompanying attributes.

Revisiting the fault tree in figure 3, it was found that the basic and initiating events that have human error events in them are B1 to B5. From table 2 it can be seen that the factors influencing the occurrence of the basic event "insufficient volume in tank" are procedures, documentation, training, checks and displays. Using equation 1 the human factors modification index β can be calculated. The performance measure r_i' is obtained by carrying out an audit. When doing an audit it is recommended that the whole human factors spectrum is covered. From such an audit it will be possible to pick any r_i'. The procedure for obtaining the performance measure value is the same as that described in [14].

The performance rating is represented in a five-point Likert scale and the excellent conditions assume 1 and poor assume 0. Average conditions will be assigned 0.5. The scale is shown in figure 6.

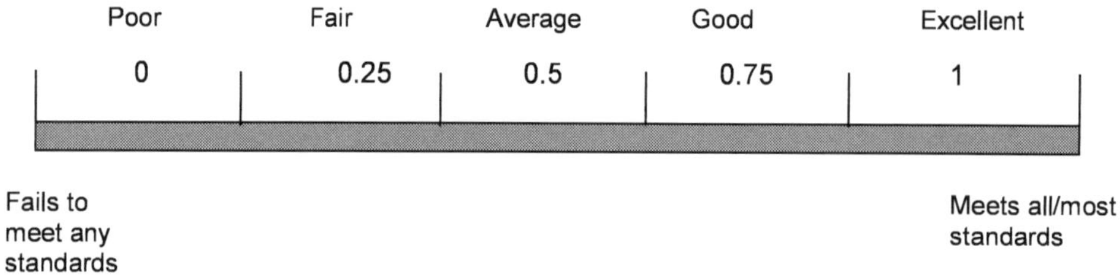

Figure 6: The scale showing human factors rating

Since the aim is to reduce the uncertainty margin to reflect the current plant conditions, the adjustment is done using the following formula:

$$HEP_{calculated} = HEP_{UB} 10^{\Theta\beta} \qquad (2)$$

where, $\Theta = \text{Log } HEP_{LB} - \text{Log } HEP_{UB}$

and HEP_{UB} and HEP_{LB} are the upper limit/bound and lower limit/bound human error probabilities respectively.

5.3 Practical Example

We will use results of an analysis done on two facilities [14] to calculate the human factors indices β for each of the two basic events B_1 and B_2. This is done using equation 1 and results are tabulated on table 3. It is seen that the factors associated with the occurrence of the human error events, B_1 and B_2 are below industry average for facility A while those for facility B are above the industry average. Industry average is obtained when the value of HF modification index $\beta = 0.5$.

Using equation 2, for both basic events $\Theta = -1.95$

Taking basic event 1,
$HEP_{calculated} = 0.0114$ for facility A and 0.0019 for facility B.

For basic event 2,
$HEP_{calculated} = 0.0223$ for facility A and 0.0091 for facility B.

Table 3: Calculation of human factors index β for two facilities

	Identified Influencing factors	Normalised weights, ω_i'	Facility A, r_i'	Facility B, r_i'
Basic event 1	Procedures	0.15	1	1
Insufficient volume in tank	Documentation	0.12	1	1
(HEP range = 0.001 – 0.09)	Training	0.23	0.5	1
	Supervision checks	0.29	0.25	0.5
	Displays	0.21	0	1
			β=0.46	β=0.86
Basic event 2	Displays	0.13	0	1
High level alarm fails or ignored	Lighting	0.06	0.25	0.25
(HEP range = 0.001 – 0.09)	Accessibility	0.14	0.25	0.75
	Attention / Motivation	0.28	0.25	0.25
	Skills and Knowledge	0.25	0.75	0.75
			β=0.31	β=0.51

Let us look at the implications of this analysis. The original FTA has used the industry mean (nominal probabilities). That is, the logarithmic mean of the probability distribution and in this particular case the cutset $B_1.B_2.B_5 = 3 \times 10^{-2}$/yr (see section 4). In our analysis the plant specific conditions have been introduced. The results have shown a very big difference between the value of the same cutset in facility A and facility B. For example, the analysis of facility A shows that the same cutset $B_1.B_2.B_5 = 8.6 \times 10^{-2}$/yr.

Had the industry average been assumed for this facility the risk could have been underestimated by a factor of close to 3. The same cutset has a value of 6 x 10-3/yr in facility B. For risk reduction strategies it is clear that displays require special attention for facility A. Also the factors that affect attention and motivation should be addressed since this attribute has a big weight.

6. Conclusion

HF were introduced into quantitative risk analysis using a newly developed procedure. It has also been argued in this work that introducing these factors will increase the accuracy of human error probabilities. In this case it means making HEPs better or worse, depending on the audit results. Here fault tree has been chosen as accident modeling technique. The unwanted event is deductively broken down into basic events and the ones that contain human error events are determined. After these events have been identified the underlying causes are investigated. A procedure to identify and analyze underlying HF causes was developed. It was clear that the underlying factors have different weightings and with the help of AHP a weighting system for these factors was developed. Combination of the weightings and the audit result for each factor was used to calculate the human factors modification index, which in turn is used to modify the HEPs. The effectiveness of this approach was demonstrated by use of a case study. It was shown that in one case the frequency of a top event was underestimated by a factor close to 3.

It is worth mentioning that the weighting system used here was obtained from results of a survey held throughout European chemical process industry. The use of representatives from the industry brought in practical sense in this work.

7. REFERENCES

1. Joschek, H.I., Risk Assessment in the Chemical Industry, *Proceeding of the International Topical meeting on Probabilistic Risk Assessment*, New York: American Nuclear Society, 1981.
2. McCafferty, D.B., Successful System Design through Integrating Engineering and Human Factors, *Process Safety Progress*, New York: American Institute of Chemical Engineers, 14(2): 147-151, 1995.
3. HSE, *Reducing Error and Influencing Behavior,* Sudbury: Health and Safety Executive, 1999.
4. Löwe, K. and Kariuki, S.G., Methods for Incorporating Human Factors during design phase, *Loss Prevention and Safety Promotion in the Process Industries*, Loss Prevention, pp. 5205-5215, 2004.
5. Vuuren, W.v., Shea, C.E. and Schaaf, T.W.v.d., The *development of an incident analysis tool for the medical field,* Report, Eindhoven: Eindhoven University of Technology, 1997.
6. Reason, J., *Human Error*, New York: Cambridge University Press, 1990.
7. Chapanis, A.R., Human-Factors Engineering. *The new Encyclopedia Britannica*, Chicago: Encyclopedia Britannica, 1986.
8. Kariuki, S.G. and Löwe, K., Increasing Human Reliability in The Chemical Process Industry Using Human Factors Techniques, *Process Safety and Environmental Protection,* 84(B3): 200 – 207, 2006.
9. Attwood, D.A., Deeb, J.M. and Danz-Reece, M.E., *Ergonomic solutions for Process Industries,* Oxford: Gulf Professional Publishing, 2004.
10. Löwe, K., Kariuki, S.G., Porcsalmy, L. and Fröhlich, B., Development and Validation of a Human Factors Engineering Guideline for Process Industries, *Loss Prevention Bulletin* (lpb)(Issue 182): 9-14, 2005.
11. PRISM, Incorporation of Human Factors in the Design Process. www.prism-network.org., 2004.
12. AIChE, *Guidelines for Chemical Process Risk Analysis*, New York: American Institute of Chemical Engineers, Center for Chemical Process Safety, 2000.
13. Saaty, T.L., *The Analytic Hierarchy Process*, New York: McGraw-Hill, 1980.
14. Kariuki, S.G., Integrating Human Factors into Quantitative Risk Analysis. *PHD Thesis*, Berlin: Technical University of Berlin, 2007.
15. Kariuki, S.G. and Löwe, K., Integrating Human Factors into Process Hazard Analysis. accepted: *Reliability Engineering & System Safety*, Elsevier, 2007.

Safety Culture: "Black Art" or "Paradigm Shift"?

Cinzia De Santis, Patrick Hudson, Matthew Lawrie, Cory Shelton, Dave Rose, Ard van Bergen, Diane Chadwick–Jones, Evelyn Mitchell
Contact email: santisc@bp.com

Abstract
Catastrophic industrial accidents have been frequently associated with breakdowns in Safety Culture. Statistical analysis performed in BP showed evidence of the relationship between Safety Culture and incidents. In June 2005 a project was started in BP to identify a Safety Culture model applicable to all activities and across the entire Group. The applicability and validity of the model was tested through assessments in 5 pilot sites and the data analysed using basic statistical analysis (median, mode and standard deviation). In parallel, best-in-class organizations were visited and an extensive review of the theory was carried out. The knowledge acquired was the basis to construct the Safety Culture model proposed for BP. The model in based on existing theory but has been adapted to BP's operations worldwide. An organization with outstanding Safety Culture needs to be: informed (leaders know what is going on and are competent), mindful (everybody thinks of what might go wrong), continuously learning (all events are seen as opportunities to learn), fair (everybody knows the rules and are held accountable for their actions) and respectful (people feel included, treated fairly and their personal circumstances are important). The model emphasizes the importance of inclusiveness and trust with respect to Safety.

Introduction

There are many different definitions of Safety Culture, among which the following two are the clearest: *"The safety culture of an organization is the product of individual and group values, attitudes, competencies, and patterns of behaviour that determine the commitment to, and the style and proficiency of an organization's health and safety management" (UK HSE, 2002);* and "*Shared values (what is important) and beliefs (how things work) that interact with an organization's structures and control systems to produce behavioral norms"* (Reason, 1998).

A strong Safety Culture *per se* is not enough to deliver an outstanding performance, but needs to be in place though to underpin Process Safety, Systems and Engineering. A weak safety culture can be evident through the actions, and inactions, of personnel at all levels of the organization no matter how robust the systems and the engineering are in that organization.

Academic research, high-risk industries and governmental agencies have identified Safety Culture as one of the key root causes of incidents. In addition, major disasters in recent history (Chernobyl, 1986; Longford Gas Plant, 2000; Columbia Space Shuttle, 2003, amongst others (UK HSE, 2002) have pointed to a poor safety culture as one of the main causes of those events. The company widely considered "best-in-class", Dupont, considers Safety Culture to be the basis of both outstanding safety and business performance (Grubbe, 2006). In summary, both research and industrial experience have shown that Safety Culture is critical to drive Safety Performance.

Levels of Safety Culture

Many efforts have been made to understand and influence Safety Culture due to its well-established relevance in the overall performance of organizations.

Every organization has a safety culture, at one level of development or another. To improve that safety culture, it is necessary to: determine the level at which the safety culture currently operates, determine the ideal level to operate, and then implement changes in order to reach that ideal level of safety culture.

Researchers studying safety culture have classified company cultures into several levels—based in part on how they deal with safety-related information. Westrum (1991) proposed a three-level classification scheme for communication flow in an organization (pathological, bureaucratic and generative). Based on this model, Hudson (2001) expanded it and proposed a 5 level continuum:

- **Pathological** Safety is regarded as a problem caused by workers, the main driver is the business and a desire not to get caught by the regulator. The organization cares less about safety than about not being caught
- **Reactive** The organization starts to take safety seriously but there is action only after an incident has taken place
- **Calculative** Safety is driven by management systems, with much collection of data. The focus is on the collection and analysis of the data, rather than learning from it
- **Proactive** There is the realisation that with improved performance the unexpected is a challenge; workforce involvement starts to move the initiative away from a purely top-down approach
- **Generative** in which safe behaviour is fully integrated into everything the organisation does, there is active participation in safety at all levels, and safety is an integral part of the business. Organisations at this level are characterized by the term "Chronic Unease"

(see Figure 1: The Safety Culture Ladder)

Understanding the Culture in BP

BP has taken several steps to strengthen its organizational Safety Culture, amongst those: implementation of behavioural safety observation programs for leaders and supervisors; implementation of Group Standards for Driving, Integrity Management and Control of Work; a more rigorous, uniform and consistent Safety and Operations Management System across the whole group; addressing the competencies of workforce; creating a robust Audit process; and emphasizing the importance of Safety in the reward system for high level leaders.

In specific relation to Safety Culture and behaviours, a study was undertaken in 2005 using statistical analysis tools to identify which were the most significant factors related to Safety performance. The data analysed was obtained from different sources within BP: tr@ction (BP global Incident Reporting engine, which included: incidents, man hours, safety observations, kms driven), Major Incident Announcements, High Potential incidents, People Assurance Survey, Countries Culture, High Reliability Organization Assessments and a Safety Culture assessment using the Oil and Gas Producers (OGP) Model. The dependent variable in this study was the overall safety performance of a Business Unit, including not only personal safety statistics but also spills, major events, material releases etc.

Amongst all factors considered in the analysis, some specific questions of the People Assurance Survey related to the quality of Leadership and Safety Culture showed the strongest significant correlation with Safety Performance. This was not only the case for incidents related to personal or behavioural safety but also for major incidents. These findings were in concordance with the established theory, and invited further enquiry.

To understand better the essence and the key components of a strong safety culture, field visits were undertaken in high-risk organizations with exceptionally few accidents and an extensive review of the literature was carried out.

Recent studies of large, formal organizations that perform complex, inherently hazardous, and highly technical tasks under conditions of tight coupling and time pressure conclude that most will fail at some point, causing very serious accidents to occur. There is a small group of organizations that appear to succeed under such testing circumstances with fewer than "normal" or expected accidents. These are sometimes termed *high reliability organizations* (HROs) (Weick

& Sutcliffe, 2001; LaPorte 1996). The signature of an HRO is not that it is error free, rather that errors do not disable it. HROs put a premium on expertise, and look for personnel with extensive experience, skills and training. Decisions are pushed down the line of command to where expertise resides.
HROs do not operate in a command and control world as this works only when there is a stable environment but not when the unexpected happens. HROs cannot afford to be wrong - an accident on an aircraft carrier or in air traffic control can be catastrophic.

In order to understand better how HROs operate, field visits included nuclear plants, flight operations aboard an air craft carrier and a chemical plant. (See Figure 2 – Photo of aircraft carrier flight deck)

With the information gathered from the pilot studies in different parts of BP, the field visits, and the theory, it was defined with more detail how the **Generative** level of the safety culture ladder would look for a company such as BP and this was considered as the way in which BP assets want to operate.

In this ideal safety culture, the organization displays the following characteristics:

Informedness:
Communications are fluid in both ways, so the workforce tells management bad news and management is willing to listen. Management knows what is going on and is prepared to provide positive and/or negative feedback. Everybody is prepared to report their errors and near-misses, and attention is paid to details and small events. Managers know what questions to ask because they are competent and they make sure their workforce is competent as well. Managers listen to those who know.

Mindfulness:
Weak signals are reported and seen as opportunities to see into the system. Information is not discarded or simplified. Everyone thinks about what *might* go wrong. People can deal with the unexpected, and they are well prepared to deal with it. People listen to those with knowledge, and decisions are delegated to the level in the organization where the expertise resides.

Learning
There is learning from incidents which is acted upon quickly. Organizations seek new ways of understanding and working, and procedures are under constant scrutiny. Training and more training creates competence, while cross-training supports effective communication between functions and specialists. There is a clear process for continuous improvement.

Fairness
Everyone knows what is acceptable, as they helped define the standard and it has been properly communicated. Everyone is aware of the consequences of their actions, both positive and negative, and communications are clear in this sense. There is not a blanket no-blame approach, but blame is reserved for truly egregious behavior, involving recklessness or malice.

Respectfulness
The workforce is involved in all aspects related to safety and they are encouraged to participate. People's ideas are listened to and taken into consideration. Workers are treated with respect and safety messages take in consideration the local culture within which the organization is operating. Workers personal circumstances are taken in consideration.

Based on this model a Safety culture assessment tool has been devised to:
1) Assess safety culture strengths and weakness in the Business Unit/site/platform,
2) Learn how perceptions vary among organizational elements and hierarchical levels (including contractors), and
3) Identify areas of opportunity for material improvements in Safety Culture

The assessment tool covers 12 elements of Safety Culture as follows:

- Management commitment
- Two-way communication
- Workforce behaviors
- Learning culture
- Production or cost vs. safety
- Management systems
- Attitudes towards procedures
- Relationships with contractors
- Local cultures
- Team leader trust and respect
- Major incidents
- Attitudes towards risk

These are comparable with commonly cited important elements of safety culture from academic and industrial literature, the regulator and include elements inspired by internal BP studies. Two to three safety culture survey questions were devised for each element as listed above. For each question, there were five responses relating to each of the levels of safety culture. Table 1 shows examples of the question and answers relating to risk assessment and risk tolerance.

Table 1: Example of questions and possible answers in the safety culture assessment tool relating to risk assessment and risk tolerance.

How are risk assessment tools used?	Risk assessment is done in isolation by the person responsible for safety.	The organisation builds risk assessment tools around the incidents that it sees.	Many risk assessment tools are used in a systematic way, with frontline staff formally involved but not really engaged.	There is an understanding of the need to practise informal risk assessments on a day-to-day basis in addition to scheduled ones.	People practise day-to-day risk assessment in both formal and informal ways to effectively watch out for each other as well as for themselves.
How tolerant are individuals of risks?	They have to tolerate the risks if they want to get paid.	They don't take chances with things they know have hurt people in the past.	They don't take chances with the things that can obviously hurt them.	They look at the job ahead and don't take chances that they think could hurt them.	They think through every activity well in advance and don't take chances with their own or others' safety.

The assessment is completed on paper or in an electronic format and in the local language(s). As well as English, the assessment tool has been translated to Azeri, Russian, Arabic, German, Dutch, Danish and Spanish. The tool is reviewed by the Business Unit, Asset or platform before starting the assessment process to make sure that the language is understandable and compatible with the specific activity of the Business Unit (e.g. Marketing, Lubricants, Upstream, etc.) The full assessment cycle consists of a 12 Step process. Each step is described by a procedure and all the procedures are available on an internal web site.

Although simply carrying out the safety culture assessment helps raise the awareness and enables a conversation around the safety culture in the organization, the most critical phase of the cycle is related to the improvement plan. Once the gaps have been identified (see Figure 4: Example of results from a safety culture assessment) the leadership team of the unit agrees upon and commits to implement the necessary changes to move to the next level of safety culture.

The improvement program is based mostly on a coaching framework where coaches are identified in the business unit and trained to support their own leaders to improve the safety culture of their organization. Best practices from other parts of BP and other organizations, training programs and a continual support from the central organization are also made available to the business units where the assessment and improvement process is taking place.

The importance of the global approach is emphasized by having a cross culture – cross business operating model.

Conclusions

The project has raised awareness of the importance of group and individual behaviours, social interactions at work, values, beliefs and the relevance of local culture in the way organizational safety is managed. In some cultures the act of challenging one's superior is more difficult than in others and challenge is crucial when safety is being compromised. The project also helped to emphasise the importance of credibility and trust in being able to deliver an outstanding culture. These are "perishable goods" within an organization, and as such they have to be continuously nurtured and renewed. Every day is a new day in forging and maintaining these relationships; it never gets institutionalized.

The assessments completed so far point toward both barriers and enablers, on a local and organizational level, to improving organisational safety. The assessment acts as a "mirror" for the organizations involved in them, enabling them to reflect in an objective way on how people at all levels in the organization perceive their safety culture. The understanding of what a safety culture is and how it is experienced are used in the assessment tool and support the coaching program. It takes such a difficult concept as an organizational culture from the realms of the 'Black Arts' into forms people can recognise in their everyday activities.

The improvement tools and coaching program based on the findings are helping business units/sites to develop and enhance their Safety Culture. Additionally, some interventions have been made at organization level to support the improvement of overall BP Safety Culture. The changes that can be introduced form a paradigm shift as the organization moves up the safety ladder.

References and Notes
Cullen, The Lord D. (1990)*The Public Enquiry into the Piper Alpha Disaster.* Her Majesty's Stationary Office, London.
Grubbe, D. (2006). Personal communication.
A Hopkins, *Lessons from Longford* (Sydney: CCH, 2000),
Hudson, P.T.W. (2001) Safety Culture: Theory and Practice In *The Human Factor in System Reliability.* Nato Series RTO-MP-032. North Atlantic Treaty Organisation: Brussels. [CD-ROM]
La Porte, T.R. (1996) High Reliability Organizations: Unlikely, Demanding and At Risk. *Journal of Contingencies and Crisis Management*, **4**, 60-71
NASA (2003) Columbia Accident Investigation Report. Houston, TX.
Reason, J. T. (1998) *Managing the Risks of Organisational Accidents.* Ashgate, Kent.
UK Health & Safety Executive, (2002) *Safety Culture: A Review of the Literature,* HSL/2002/25.
Weick, Karl E. & Sutcliffe, Kathleen M., (2001) Managing the Unexpected: Assuring High Performance in an Age of Complexity, Jossey-Bass.
Westrum, R.J. (1991) Cultures with Requisite Imagination. In J Wise, P Stager & J Hopkin (Eds), Verification and validation in complex man-machine systems. New York: Springer.

List of Figures

Safety Culture: "Black Art" or "Paradigm Shift"?

List of Figures

1. The Safety Culture Ladder

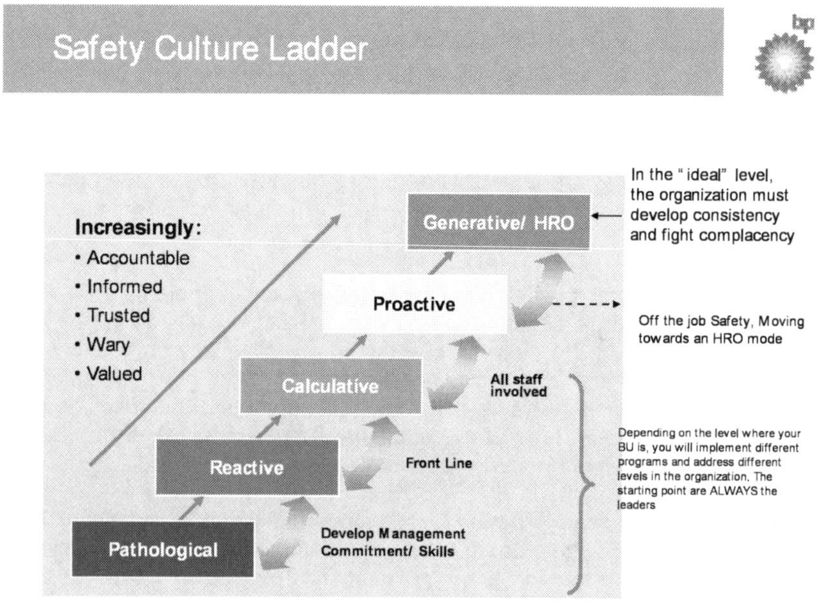

2. Photo of aircraft carrier flight deck

3. Example of results from a safety culture assessment

Example Safety Culture Profile BU / Site

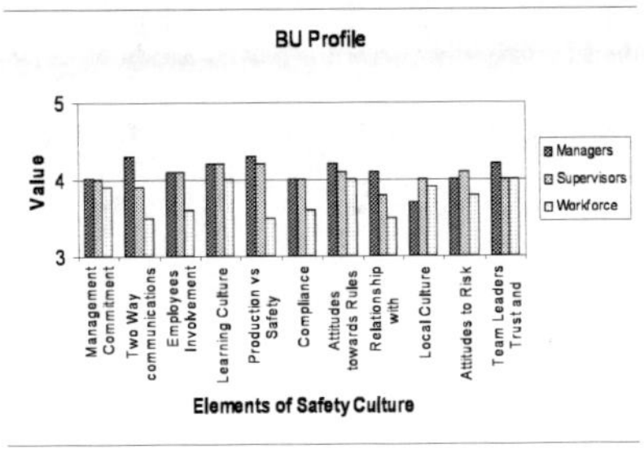

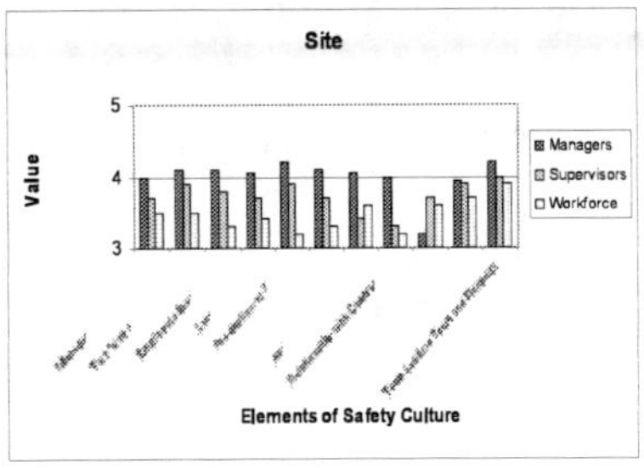

- Assessment of risk profile, discriminates between sites. Acceptable level: 3.5
- 11 Elements of Safety Culture
- Regional Focus
- Assessments identify strengths/ weaknesses, interfaces between groups (horizontal) and between levels (vertical) in a same BU/PU
- Cross – referenced with other data source: r@ction and HiPo/ MIA and other data base (country culture)
- Focus groups to identify root causes and improvement plan

Six Sigma®[1] Analysis Applied to Process Safety Systems

Jon C. Ritz, Process Safety Manager, The Lubrizol Corporation

Brecksville, Ohio

Jon.ritz@noveon.com

Jon A. Bernardi, Responsible Care® Systems Manager, The Lubrizol Corporation

Brecksville, Ohio

Jon.bernardi@noveon.com

ABSTRACT

Process safety management (PSM) is about reducing the number and severity of process safety incidents. We sought to apply the Six Sigma concept of continuous improvement to the area of process safety. To strive for further improvement in process safety we needed to methodically delve more deeply into understanding the underlying issues and causes of incidents and to look beyond high profile events that are, fortunately, quite rare. We brought together a team from multiple sites to develop a standard set of criteria for incidents and looked at historical data from a select number of plants to develop our base line. We refined the incident criteria and implemented the data collection process globally. The first look at the Six Sigma analyses helped us to further hone the definitions. Ultimately, this led us to use the major elements of the OSHA PSM Standard 1910.119 for the analysis, to give us a standard reference. From there, we identified two major focus areas (mechanical integrity and operating procedures/operator training), where we are implementing the second level of follow up to update global standards and approaches.

The presentation will focus on the detail of the methodology leading to the identification of mechanical integrity and procedures/training as the two major focus areas, lessons learned, overall feedback and next steps. As an added benefit, by collecting more detailed data, we identified various site issues: rupture disc release and hose handling, which were handled at the specific sites

1. Company Background and Six Sigma Methodology

Noveon, Lubrizol

Noveon, a wholly owned subsidiary of The Lubrizol Corporation, manufactures a variety of specialty chemicals and plastics and has its headquarters in Brecksville, Ohio. The Lubrizol Corporation is a worldwide manufacturer of additives for engine oil, fuels and lubricants. Lubrizol's headquarters is in Wickliffe, Ohio.

---

[1] Six Sigma is a trademark of Motorola Inc. for a system of practices for continuous process improvement

Six Sigma Methodology - Define, Measure, Analyze, Improve and Control

Six Sigma is a proven methodology for continuous process improvement. It is based on a rigorous data-driven approach, following the steps of define, measure, analyze, improve and control. The Noveon subsidiary has been using the techniques in manufacturing, marketing, R&D and finance since 2003. One distinction with this project is its extended time frame. Usually a Six Sigma project is completed in four to six months. This project represents a long-term cultural change. The collection of data and the need to understand the incident potential make this an ongoing task.

2. Project Charter

Improve Process Safety Performance

Historical data, as measured by the overall American Chemistry Council (ACC) criteria, showed that we were in statistical control but that the amount of variability in process safety performance was not acceptable.

Define Lower Level Metric (Below ACC Criteria)

Along with performance improvement, we needed to develop a metric for process safety that would allow us to track progress at a lower level of severity.

3. Approach (Define)

We selected team members from a number of U.S. plants. We needed their expertise, experience and knowledge of items that were being tracked in their facilities to help us develop corporate guidelines.

We met and asked each member to bring the data he/she had accumulated. We put this into one spreadsheet to understand similarities, differences, and definitions. Then we looked for a reasonable set of categories to group the incidents. These are listed in the table below:

Incident Categories

Any release, spill, emission or discharge event that impacts the community. An impact is classified as a response by the community to the event, such as a communication from neighbors complaining about the incident or the community takes some type of emergency response. For example, this would not include non-hazardous material contained in a concrete dike or building.
Any release, spill, unplanned emission or discharge that results in any employee, contractor, visitor or neighbor being admitted to a hospital.
Any release, spill, unplanned emission or discharge that results in the activation of the plant emergency response plan.
Any release, spill, unplanned emission or discharge that requires the facility to communicate to any (local, state, provincial or federal) government authority that the event has occurred. This would include precautionary calls that are later determined not reportable.

The activation of any relief valve, relief valve/rupture disc, or rupture disc resulting in a spill, release, unplanned emission, or discharge.
Any fires or explosions at the facility.
The activation of a safety interlock that results in an unplanned emission, spill, release or discharge.
Any unplanned environmental control device related event with potential for outside agency follow up.

Then, we further defined categories of incidents as follows:

Class	Definition
Failure	Did a piece of equipment, device, pipe or valve fail and cause the incident? Other than the specific items listed below.
MisOp/Error	Was the incident primarily due to an error or mis-operation of some type?
RD/RV	Was the primary incident due to the activation of a rupture disk or relief valve? If an RD activated due to an error, classify the incident still as RD/RV.
Leak	Was the incident due to a leaking flange, valve, or gasket?
Fire	Was the incident due to a fire, decomposition, or violent reaction?
Overfill	Was the incident due to material overflowing a tank?
Control Device	Did some PSM-related control device malfunction and cause an incident?
Hose	Did a hose rupture, fail or otherwise cause the incident?
Other	Anything else (to be used rarely!)

Then we began the analysis work with the data from the sites represented on the team. At this point, we had a workable set of definitions and classifications and rolled the data collection out to the entire corporation.

4. Project and Data Analysis (Measure and Analyze)

We began the data collection against the new criteria at the beginning of 2004. Normally in a Six Sigma project, this phase of the project should be complete in a relatively short time. Fortunately, process safety incident data does not accumulate rapidly, so the first phase of the data collection lasted approximately one year. We did begin looking at the data in the third quarter, to determine if we could pick out trends. This is shown in the following chart:

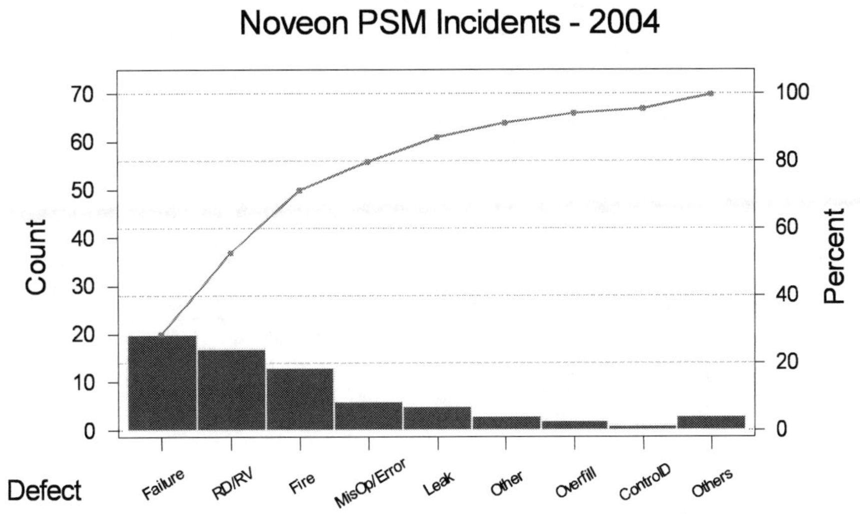

Defect	Failure	RD/RV	Fire	MisOp/Error	Leak	Other	Overfill	ControlD	Others
Count	20	17	13	6	5	3	2	1	3
Percent	28.6	24.3	18.6	8.6	7.1	4.3	2.9	1.4	4.3
Cum %	28.6	52.9	71.4	80.0	87.1	91.4	94.3	95.7	100.0

However, we discovered that the new categories or classifications were not as clear as we would have liked. For example, if there were a release or spill, and it was due to a rupture disk blowing, how do we classify the incident? The classifications were too focused on the event outcome instead of the system failure.

We established a severity rating grid to determine if there were any other patterns that emerged from the data. We could glean no new information. We decided we did not want to impose a further criterion on the analysis and potentially diminish the number of incidents. We did put a quantitative measure of "Serious – Yes/No?" in place, but only for use as a general guideline.

At this stage, we stepped back and took another look at the classifications. We kept the definitions for incidents the same, but we decided that a better method of classification was to use the elements of OSHA 1910.119. This was a more concise approach, and the analysis was more fruitful. This is shown in the chart below:

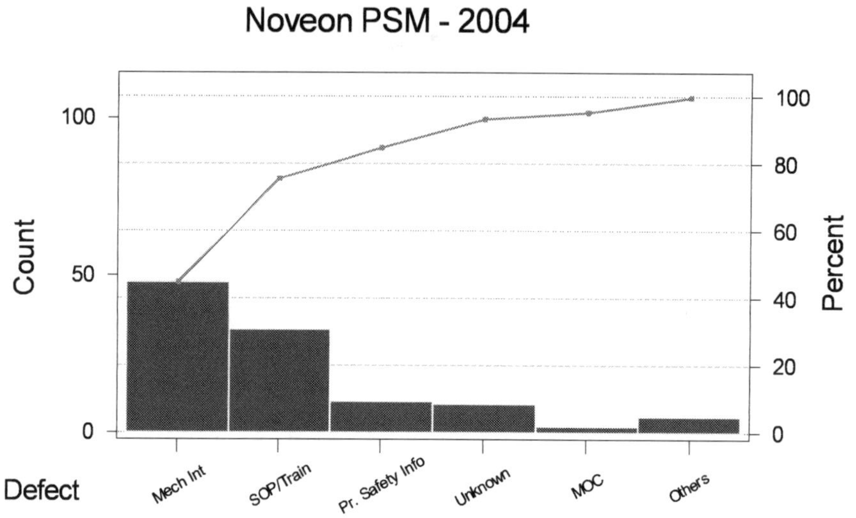

Defect	Mech Int	SOP/Train	Pr. Safety Info	Unknown	MOC	Others
Count	48	33	10	9	2	5
Percent	44.9	30.8	9.3	8.4	1.9	4.7
Cum %	44.9	75.7	85.0	93.5	95.3	100.0

The combination of Mechanical Integrity and Procedures/Training made up almost 80% of the recorded incidents.

At this stage, we felt we had an actionable direction and proposed the formation of two work teams for a more focused improvement effort.

> Mechanical Integrity (from the maintenance areas)
> Standard Operating Procedures/Training (from the operations and training functions of our sites)

We continued to collect the data.

5. Address the Issue (Improve and Control)

Then we ran into resource issues and decided that we would begin the effort with the Mechanical Integrity team. We chartered a team and began looking at further metrics. This turned out to be more of a challenge than we realized because there is no standardized maintenance system across the corporation. We did bring a group together made up of representatives from facility maintenance groups in the U.S. and Europe. This group was charged with developing a new corporate policy (Standard of Care). This policy includes eight guidance documents related to specific elements of mechanical integrity, vessels, pumps, control systems, etc., to further support the detail needed to manage the Mechanical Integrity process. The collection of data was completed with email and a central collection point. This has now been shifted to a corporate intranet system, where each site can input their incident data directly.

6. Benefit

The benefit we are now beginning to realize is a standardized operational approach. This is due partly to the new Standard of Care for Mechanical Integrity and the fact that we are using common definitions and reporting tools. There is now a single basis for the way we handle mechanical integrity for the company, while allowing for site-specific differences due to local issues, plant scale, etc.. We also are able to identify site-specific improvements. For example, when reporting the data one of our sites noticed a number of rupture disc incidents, which upon further investigation were due to unclear and poorly understood operating conditions and procedures. These have been addressed, and the rupture disc failures have been almost eliminated. Another site uncovered a pattern of failure with flexible hoses. They found a better style of hose, increased their inspection frequency, and upgraded their site hose handling policy.

Based on results of our study, we determined that in addition to the need to establish corporate metrics for monitoring Mechanical Integrity performance we should allow each location to define local metrics based on their specific issues.

7. Road Blocks

We continue to struggle with several issues. Data acquisition, while greatly improved, needs constant reinforcement. Another issue is data integrity, making sure the data are in the correct category, there is enough supporting detail, etc. We are working toward a standardized action item data capture software but are in the initial stages of that project.

8. Changes and Next Steps

During the course of the project, Noveon has seen many changes. We went from being a privately held company to being acquired by Lubrizol. Then we had a divestiture of non-core businesses. But through it all, the data and trends seen remained constant, giving us the belief that the core issues were correct and that we are working on the right things and in the right direction.

We are just beginning the second leg of our effort, focusing on Operating Procedures/Training. Again, we will use a cross-company group, but this time we are using people from the sites directly involved in training and procedures.

We continue to monitor and analyze the process safety data.

9. Six Sigma Application

A final question comes down to, "Why use the Six Sigma approach?" Six Sigma provides the framework to go after an issue using a data-driven methodology, even in areas where hard data have traditionally been difficult to obtain. This forced us to develop reasonable metrics that could be analyzed and tracked. Then, once we have data, we can make and develop fact-based solutions. The major deviation we saw with the process was the length of time involved. In a traditional Six Sigma project, the time frame from start to finish is usually in the five to six month range. The difficulty we had of collecting a meaningful amount of data, and the cultural issues involved, led us to a long-term approach.

So what should you do? We would make two recommendations. First, look at your mechanical integrity and operating procedures/training. We believe you can make a major improvement by using these focus areas. Second, look at your own data. Due to the nature of your business, facility, culture, etc., there may be other factors at work. Six Sigma can help you uncover your specific targets.

Labor Accidents and the Relation with Human Factors : Case: Braskem – Raw Material Unit – Bahia. 2006. 148 f. **Master Graduation Work,** – Administration Scholl, Federal University of Bahia, Salvador, 2006

By: Américo Diniz CARVALHO NETO, Braskem

ABSTRACT

Statistical data disclosed by private and governmental institutions have pointed out the high and preoccupying rate of work accidents in industrial installations, affecting several governmental, civil society and private enterprise sectors. In the first case, the high costs of workers compensations and injury recovery expenses compromise, many times, funds to be applied in preventive actions and in workers' social welfare. Secondly, accidents leading to non-fatal injuries, and, in some cases, fatal injuries leading to death bring the family irreversible losses. At last, private initiative has lost a great deal of investments on its workers, apart from losses in terms of installations and processes resulting from accidents. Despite technological evolution observed in system and equipment components, resulting from investments in terms of reliable and high performance systems, in spite of their being automatized and capable of minimizing and even eliminating operational failures, the need for man's interface adjustment remains. In this sense, unfortunately, equivalent advances have not been observed in terms of research and tools for minimizing the probability of human failure in systems and equipment operation which can preserve adequate work conditions and human values in work environments. In this work, an assessment is made of the influence of human failure in accident occurrences in Braskem's Raw Material Basic Unit, and a methodology for human failure assessment and tools for risk and accident analysis are proposed. Human factors involved in project creation are also to be considered, so as to enable systems and equipment to be projected in safer and more reliable ways, aiming at decreasing the probability of accidents involving people, installations and environment. It is expected that this research might contribute to fostering workers' going back home as they came to work, i.e., with their physical, psychological and emotional integrity preserved.

Keywords: Ergonomy; Work safety – Human factors; Reliability (Engineering)

The Next Generation Operational Discipline Program at DuPont

James A. Klein and Bruce K. Vaughen

E. I. du Pont de Nemours and Company
Wilmington, DE, USA
james.a.klein@usa.dupont.com

The purpose of this paper is to review the development and implementation of a next generation approach to improving operational discipline programs at DuPont, in order to help sustain high levels of process safety management performance. DuPont defines operational discipline as a deeply rooted dedication and commitment by each member of the organization to carry out each task, the right way, each time. Eleven organizational characteristics are currently used to help evaluate and drive continuous improvement of operational discipline. Ultimately, though, it is individuals who do the day-to-day work required by PSM systems, and it is these actions that help determine if the organization is able to operate with high levels of operational discipline. With this in mind, DuPont's program has been revised to include a new emphasis on individual characteristics of operational discipline, based on knowledge, commitment, and awareness. Discussion will include why operational discipline is important, how operational discipline can be characterized for both organizations and individuals, and how the revised program is being implemented.

1. Introduction

Operational discipline (OD) is defined by DuPont as the deeply rooted dedication and commitment by every member of an organization to carry out each task, the right way, each time [1]. OD reflects the strength of an organization's safety culture in making Safety, Health, and Environmental (SHE) systems effective and in providing tangible results for preventing injuries and incidents. In addition, a high level of OD typically contributes to improved operational and business performance through higher productivity, higher quality, reduced waste, and lower costs. "Discipline" in this context does not refer to punishment, but to doing things in a way that results in positive benefits.

The OD program at DuPont is being revised to introduce a new focus on individual OD as well as organizational OD. This recognizes that, while organizational factors are essential for achieving high levels of OD, tasks are ultimately completed by individuals who have the knowledge, commitment, and awareness to complete the tasks correctly and safely every time. Aspects of the OD program being worked on include: new training, new guidance on evaluating and improving OD, and resources to assist implementation at sites. The OD program is meant to be flexible, providing a toolbox of methods for evaluating and improving OD, based on individual site needs.

1.1 Operational Discipline at DuPont

The emphasis on OD at DuPont is found in the early development of safety culture in the company, where the manufacture of gunpowder was the first hazardous process. In 1808, the founder, E. I. du Pont, wrote:

> The safety of our family, the safety of the farmers who live in our neighborhood, has imposed upon us the absolute duty of making choice of steady, sober men and of establishing the most rigid discipline among our workmen [2].

Nearly 200 years later, OD is viewed as essential to all SHE programs, but is most associated with Process Safety Management (PSM) systems, designed to identify, evaluate, and control a variety of potentially hazardous processes. OD is recognized as essential to the development and implementation of effective PSM systems as one of four required steps [3, 4]:

- Step 1 – Establishing the Safety Culture: A PSM program will only be as effective as the underlying safety culture permits. If an organization has a weak safety culture, PSM efforts are harmed by conflicting priorities, such as production needs or lowering costs. A strong safety culture, based on a commitment to SHE as a core value, and as evidenced by organizational policies, goals, metrics, and day-to-day decision-making, supports establishing safety systems where safety priorities are recognized, not as conflicting with other priorities, but rather as inherently necessary for completing any task the right way.
- Step 2 – Providing Management Leadership and Commitment: Management leadership and commitment which is visible throughout the organization is required for continued excellence in PSM. Beginning with the formation and fostering of a strong safety culture, continuing with policies that set high expectations for excellent safety performance, and implemented in day-to-day decisions that provide resources and establish high priority for safety-related activities, the actions of all levels of management must support and reinforce strong PSM programs and accountability.
- Step 3 – Implementing a Comprehensive PSRM Program: A comprehensive PSM program includes integrated managing systems for various elements of process safety [3-5], such as Process Hazards Analysis, Incident Investigation, and Mechanical Integrity. Regulations should generally be considered the starting point only, representing minimum essential levels of practice. Additional requirements should be considered that go beyond regulations and better meet corporate needs for process safety [6].
- Step 4 – Achieving Operating Excellence through Operational Discipline: Operational discipline is the deeply rooted dedication and commitment by every member of an organization to carry out each task the right way each time. No matter how comprehensive and well-designed PSM programs may be, it is the day-to-day ability of everyone to practice operational discipline that successfully

changes safety systems from concept to reality. Operational discipline not only helps prevent serious injuries and incidents, but also contributes to excellent business performance through increased operating excellence, including higher productivity, higher quality, reduced waste, and lower costs.

The principles and essential features of DuPont's PSM program are described by 14 elements grouped by Technology, Personnel, and Facilities, as shown in Figure 1. This figure is typically called the PSM Wheel, with each of the 14 elements, such as Process Hazards Analysis, Operating Procedures and Safe Work Practices, and Personnel Training and Performance, arranged around the spokes of the wheel. Management leadership and commitment, necessary for implementing and maintaining strong PSM programs, is shown at the center of the PSM Wheel. Operational discipline is shown as the rim of the PSM Wheel, connecting all of the elements and translating the required managing systems into real results for preventing injuries and incidents. In addition, the DuPont PSM Standard [7] includes the following requirement:

> Sites shall implement and maintain programs and systems to achieve and maintain a high degree of operational discipline for all PSM elements in a manner that supports business and operating objectives.

Figure 1 – DuPont PSM Wheel

1.2 Importance of Operational Discipline

OD is fundamental to achieving strong SHE performance. This can perhaps best be shown through the "Safety Triangle," as shown in Figure 2, where the base of the triangle represents potentially numerous OD problems that ultimately lead to poor performance in terms of near-miss incidents and potentially serious injuries. By attacking the base of the Safety Triangle to improve OD, incidents and injuries can be avoided and SHE performance can be improved. Similar figures could also be constructed that show similar behavior for environmental releases, fire safety, electrical safety, etc., including other top events such as environmental harm or business loss.

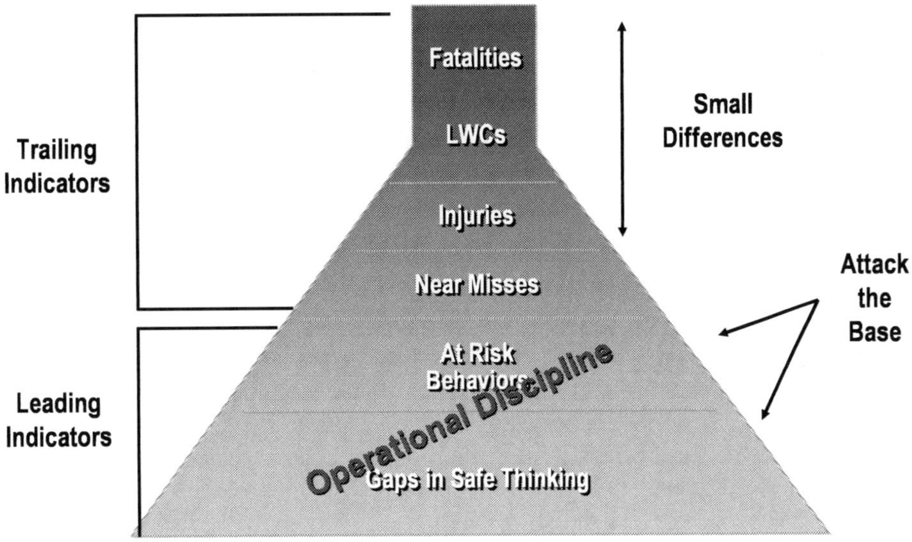

Figure 2 – Safety Triangle

A useful way of looking at OD for PSM is to modify the familiar risk equation, as shown below:

$$\text{Risk} = \frac{\text{Frequency} \times \text{Consequence}}{\text{Operational Discipline}}$$

In this equation, risk is a function of the frequency of a possible event, often determined by the effectiveness of PSM systems and safeguards, and the potential consequences of the event, characterized by the inherent substance and process hazards. OD, though, is shown in the denominator, such that if OD is 100%, meaning that the PSM systems and safeguards are maintained and operated perfectly all the time, then risk is not affected. If OD is less than 100%, meaning that even well-designed systems are not implemented or followed effectively, then overall risk is increased. If OD is only 50%, for example, then risk is qualitatively doubled. The result of poor OD and higher risk can be serious incidents and injuries, as evidenced, for example, by the Bhopal incident in 1984 [8].

2. A Next Generation Operational Discipline Program

DuPont has included OD as part of PSM training since 1989. Organizations that have a high level of OD usually exhibit an identifiable set of positive behaviors or characteristics. While it is possible that these characteristics may vary from one organization to another, DuPont introduced the following 11 characteristics in 1998 and has used them evaluate and improve OD [1]:

- Leadership by example
- Sufficient and capable resources
- Employee involvement
- Active lines of communication
- Strong teamwork
- Common shared values
- Up-to-date documentation
- Practice consistent with procedures
- Absence of shortcuts
- Excellent housekeeping
- Pride in the organization

When the effort to improve OD was undertaken, it became clear that these 11 OD characteristics were primarily the responsibility of site leadership and therefore represented "organizational" OD characteristics. Since job tasks are the responsibility of individuals, a new focus on "individual" OD seemed appropriate. As shown in Figure 3, the definition of OD has been split into both organizational and individual views. The definition of organizational OD shown in Figure 3, while important, is not what individuals likely consider when they go about their own work, perhaps believing that organizational issues are handled by someone else. Rather, a focus on conducting their own work safely is clearly in their own self-interest and control. Ultimately, it is believed that a new focus on individuals as part of the OD program will resonate better and more broadly with employees and help contribute to overall improved OD and SHE performance, as shown in Figure 4.

Organizational View

The deeply rooted dedication and commitment by every member of an organization to carry out each task, the right way, each time.

+

Individual View (New)

I am committed to working safely by doing every task, the right way, every time.

Figure 3 – Two Definitions of OD

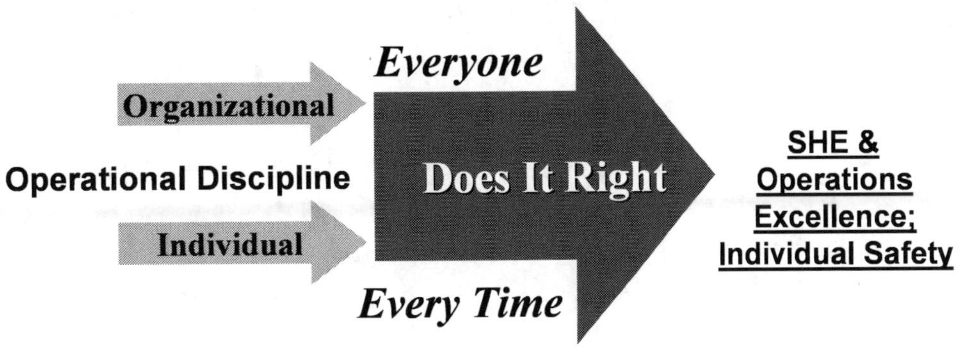

Figure 4 – DuPont OD Program

2.1 Organizational Operational Discipline

The 11 organizational OD characteristics listed above have been described in detail previously [1]. However, as the next generation OD program evolved, overlap between these characteristics became evident, and the 11 characteristics were simplified. As shown in Figure 5, most requirements from the existing program remain in the simplified version, due to the overlap and due to leadership accountability for most organizational activities. This has resulted in easier communication, training, and understanding of the program. In addition, the complexity and time commitment for evaluating OD as well as for selecting improvement practices has been reduced and simplified.

The 4 organizational OD characteristics, as shown in Figure 5, are:

- **Leadership Focus** – Leaders are passionate for SHE and model the behavior they expect from others.
- **Employee Involvement** – Employees are active and enthusiastic about participating in SHE activities.
- **Practice Consistent With Procedures** – Work is completed as planned, following authorized procedures.
- **Excellent Housekeeping** – Employees are proud of their workplace, maintaining consistently high levels of housekeeping.

The relationship between these 4 characteristics is shown in Figure 6. The foundation of any organizational effort is Leadership Focus, without which priority and support for the program cannot exist. Organizational OD also requires strong Employee Involvement, where high levels of uninterested and uninvolved employees will limit the organization's ability to achieve strong SHE performance. The most visible results of Leadership Focus and Employee Involvement are employees following approved procedures (Practice Consistent With Procedure) and maintaining equipment and work areas in good operating condition (Excellent Housekeeping). Without the base of the pyramid being in good shape, it is unlikely that an organization can consistently achieve strong performance at the top of the pyramid.

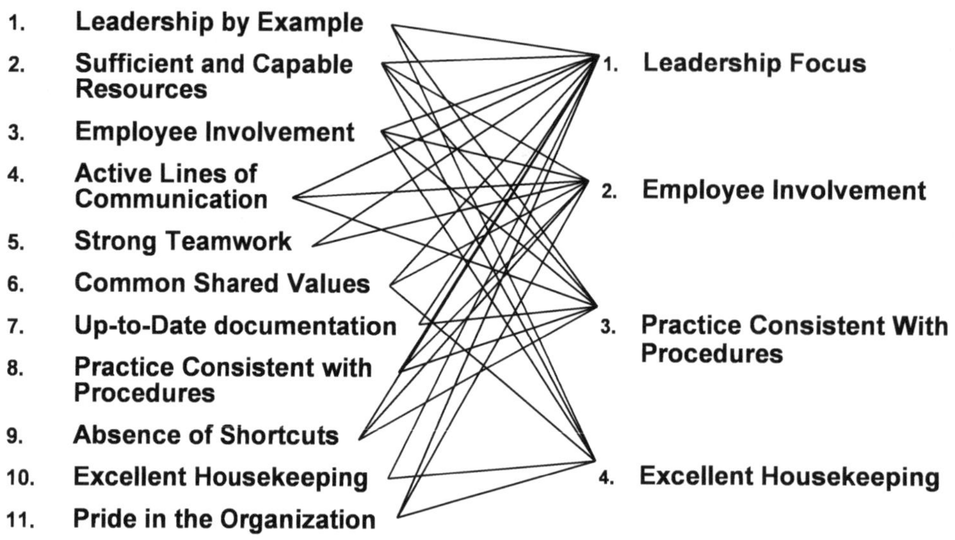

Figure 5 – Organizational OD Characteristics

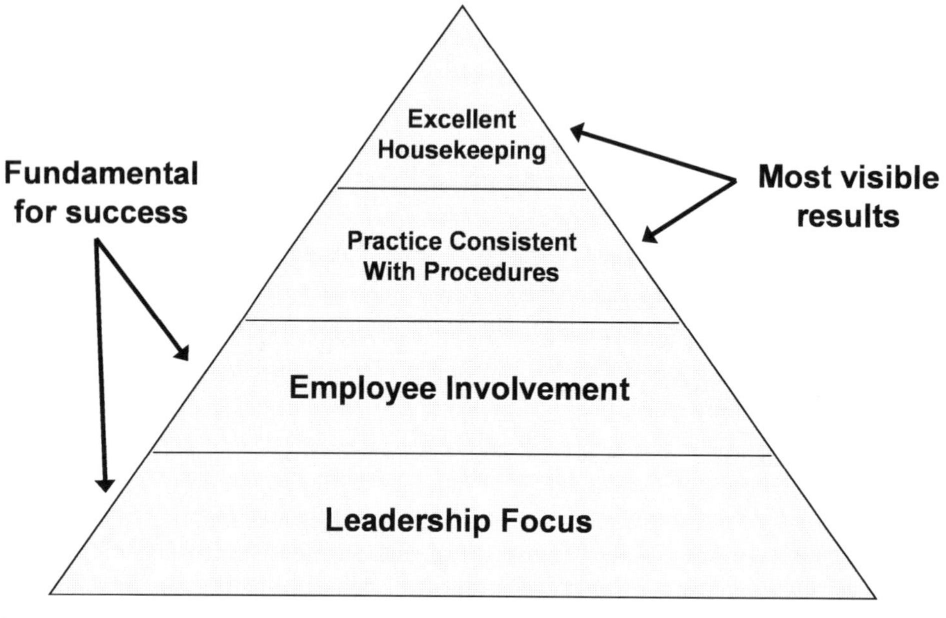

Figure 6 – Organizational OD Pyramid

Each organizational OD characteristic consists of several specific requirements, which are used to define responsibilities, to help with evaluation of OD performance, and to assist with identification of improvement priorities. The requirements for Practice Consistent With Procedure, for example, include:

- Procedures are documented and readily available for all appropriate SHE activities
- Clear expectations exist for following procedures and for not taking shortcuts
- Procedures are periodically reviewed and authorized to keep them current, including employee participation
- All changes, tests, and deviations are reviewed and authorized before use
- Training and field audits are conducted to ensure procedures are understood and followed.

Clearly, some of these requirements also reflect requirements for related PSM systems, such as Operating Procedures and Personnel Training. For OD, a key is that each requirement is carried out, the right way, every time there is a need and that organizational support for this is provided.

2.2 Individual Operational Discipline

While ultimately line management is accountable for all SHE programs at DuPont, achieving high levels of OD and subsequently achieving excellent SHE performance requires the active involvement of all employees in completing every job task correctly and safely every time. Once the concept of individual, versus organizational, OD was developed, a study was made to consider what characteristics best defined individual OD. Specific targets for these characteristics included that they should be:

- Acceptable and meaningful for individuals at all levels of the organization
- Simple and logical
- Measurable and improvable.

A broad search of related literature was conducted [e.g., 9-16] to identify key principles related to individual performance. In particular, people-based (behavior-based) safety references [e.g., 17-26] were helpful and clearly established the need for commitment as one of the required individual OD characteristics. Following extensive reading for background in this area, but not necessarily becoming an expert in any, led to a relatively simple identification of 3 characteristics required for individuals to work with a high level of OD. While some organizations may find value in a different set of characteristics, the 3 characteristics, each in first person, are:

- **Knowledge** – I understand how to do a task correctly and safely.
- **Commitment** – I commit to do tasks the right way, every time.
- **Awareness** – I anticipate potential problems and recognize unusual situations.

As shown in Figure 7, an overlap of the 3 individual OD characteristics is considered essential for an individual to work with a high level of OD. If, for example, an employee has Knowledge and Commitment, but not a high level of Awareness, when things don't go quite as planned, it is more likely that some part of the task may not be completed safely, possibly leading to an injury or incident. Similarly, if an individual is not committed to completing a task correctly and safely every time, shortcuts may be taken or distractions may occur that lead to a lack of focus. Ideally, sufficient levels of each characteristic are present, such that the employee is working in the "sweet spot" in the center of the overlapping circles.

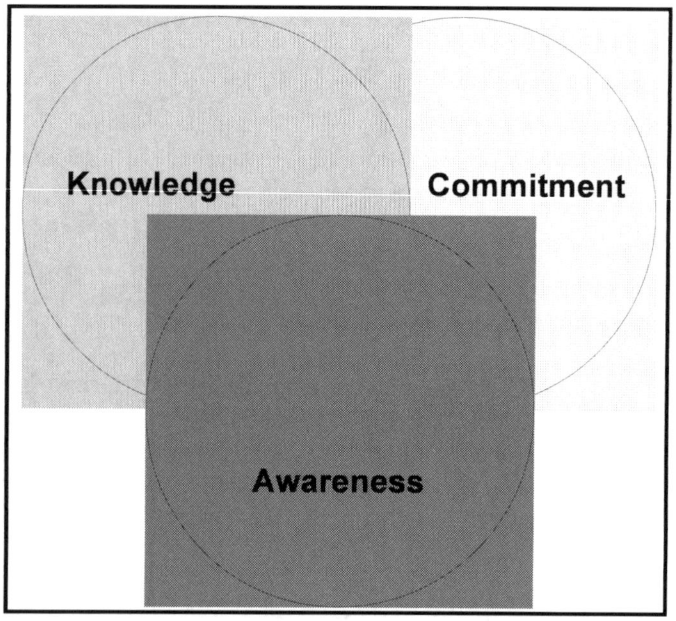

Figure 7 – Individual OD Characteristics

Each characteristic has several requirements, again in first person, which are necessary to achieve high levels of OD, building on basic tenets of safety. For Knowledge, the requirements include:

- I know the correct way to do my job task, based on procedures, training, and other SHE system requirements.
- I understand why the job task is being done in a certain way, what needs to be accomplished, and how it should be done.
- I ensure that my equipment, tools, and PPE, if needed, are in good condition.
- I make sure my co-workers also know how to do their job task safely.

For Commitment, the requirements include:

- I take personal responsibility for properly understanding my job task and making sure it can be completed safely every time.
- I plan to follow procedures carefully without shortcuts.
- I trust that procedures have been developed for a purpose, but suggest changes if they don't make sense.
- I focus on the task at hand and set aside personal and work distractions.
- I care for my safety and the safety of my co-workers.

For Awareness, the requirements include:

- I anticipate that my job task may not go as planned.
- I understand if there are any unusual circumstances or hazards associated with my job task.
- I monitor my work environment carefully, including the activity of my co-workers.
- I prepare to respond quickly and safely, troubleshooting as needed, based on my training and procedures.
- I stop, think, and review procedures, or ask for help, when unsure of what to do, or if the task unexpectedly changes.

While the specifics of Knowledge, Commitment, and Awareness may vary somewhat for different individuals in different roles in the organization, all should be meaningful in terms of completing their job tasks safely and correctly every time. Without all three characteristics, it is unlikely that high levels of OD can be achieved. While an injury may not necessarily occur if one characteristic is missing, depending on the job task, some loss of OD or operational efficiency is often a likely result. Ultimately, it is also these individual actions, based on Knowledge, Commitment, and Awareness, that contribute to the success of the organizational OD characteristics as well.

3. Program Implementation

The next generation OD program was communicated to regional groups early in its development to introduce planned upgrades and to allow for broad feedback on the program content and direction. Since different sites have different issues and priorities, it was emphasized that use of many aspects of the program would be voluntary, based on specific site needs. Specific questions asked for feedback included:

- Do you see this approach adding value for your region or site?
- How can this approach be improved to add more value?
- Do you have any input on practices that can be used to help improve OD?

Several path forward items were also identified that were being developed in support of the OD program, including:

- Improved OD evaluation tools, including auditing and incident investigation
- New guidance for improving OD
- Upgraded training.

An overview of these items is provided in this section, which are also discussed in more detail separately [27].

3.1 Evaluating Operational Discipline

The most readily available data for evaluating OD is found in site SHE metrics. Periodic review of these metrics, with emphasis on specific identification of OD contributions, can provide an excellent view of OD performance without much additional effort. Examples of metrics that may relate to OD include:

- No. of incidents (especially incidents with OD key factor)
- No. of injuries (especially injuries with OD key factor)
- No. of audit findings (especially findings that relate to OD)
- No. of scheduled activities not completed on schedule
- No. of overdue findings, tests and inspections, etc.
- Amount of waste generated
- Poor product quality and customer complaints
- Higher costs
- Lower process uptime or utilization
- Staff turnover rates

An example for the possible OD contribution to incidents is shown in Figure 8. In this scenario, it appears that OD improved in 2004, possibly due to an emphasis on OD at the site or perhaps for other reasons. Since that time, the contribution of OD to incidents at the site has steadily increased until almost all incidents have OD as a key factor or root cause. This suggests that any gains from earlier efforts were not sustained and that the site could benefit from additional focus on OD. Additional analysis of the incident data may help identify which specific OD characteristics need to be improved. Additional evaluation tools can also be used to further evaluate OD at the site.

The most useful OD evaluation tool is the OD Self-assessment [1]. The OD Self-assessment has been revised and simplified using an internet survey tool [28], which allows for rapid analysis of OD based on a survey of all site employees or subgroups, as desired. The survey collects specific responses to various graded (1 to 5) or yes/no questions, which correspond to the seven organizational and individual OD characteristics. Questions also allow participants to identify specific problem areas, to provide suggestions for improvement, or to identify current activities that seem to be working well. An example question and typical result is shown in Figure 9. Advantages of the survey tool versus previous small group discussion approaches [1] include involvement of more people, easier data analysis, and direct evaluation of participant attitudes for individual OD characteristics. If desired, the survey data can also be evaluated using a smaller group to help determine a specific, prioritized path forward.

Additional methods to evaluate OD are discussed separately [27].

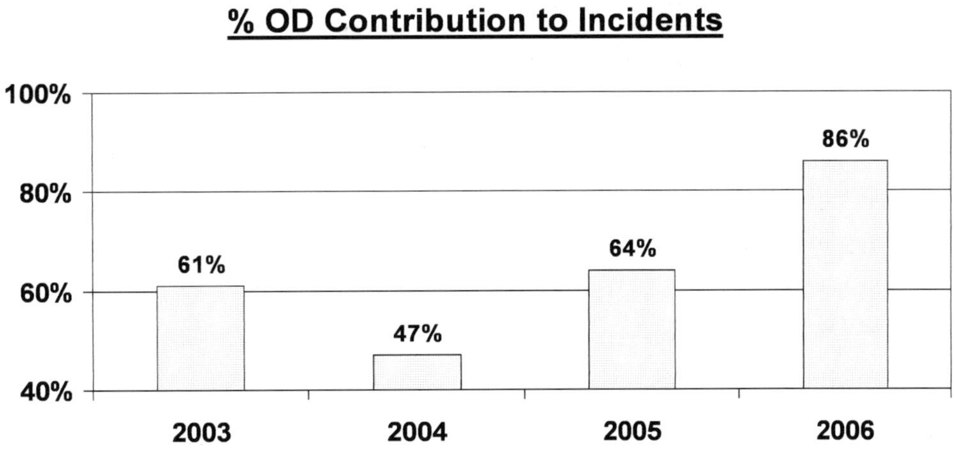

Figure 8 – Example OD Metrics Data

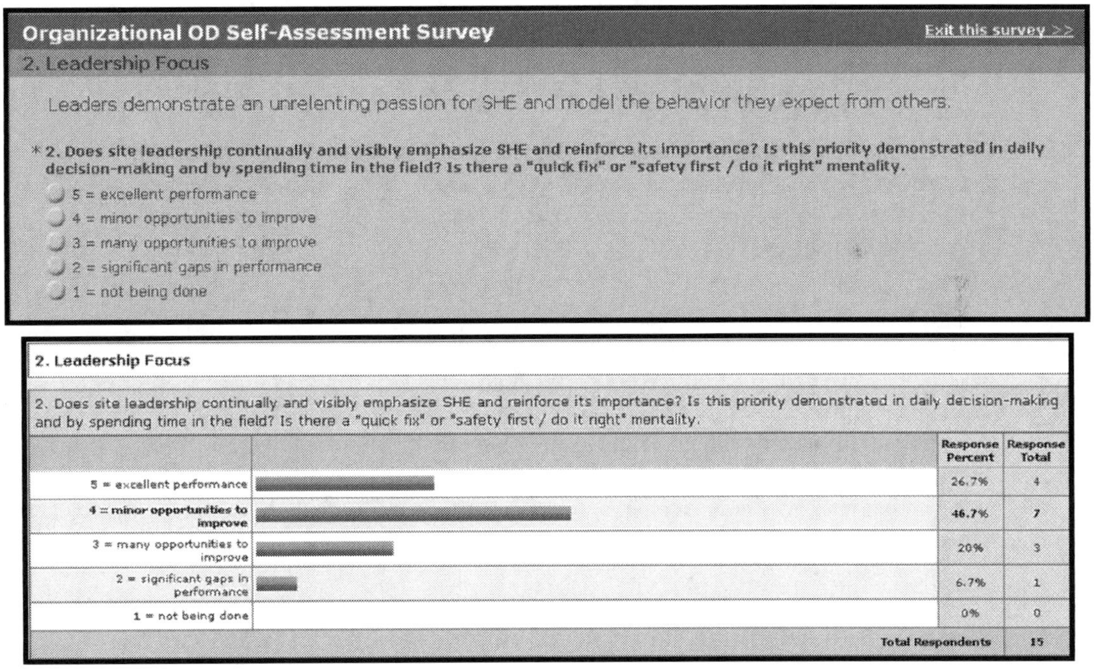

Figure 9 – Example OD Self-Assessment Survey Data

3.2 Improving Operational Discipline

The lack of specific improvement guidance for OD has sometimes limited the effectiveness of OD programs. For example, using available OD evaluation tools [1], it was possible to identify OD characteristics that needed to be improved, but when site management looked for guidance on possible approaches, little guidance was often available. In addition, improvement tools that might work at one site might not work at another, due to differing site processes and culture. An important aspect of the next generation OD program, therefore, has been to develop an internet-based "OD Toolbox" based on internal and external practices that could provide easily-accessed, flexible improvement guidance and tools. Existing practices have been combined with new approaches solicited from sites and identified from literature searches. Specific tools can be flexibly selected to target improvement in weaker OD characteristics, based on site needs and preferences. Overall, development of the OD Toolbox will be evergreen, as sites gain more experience in working on OD improvement and as additional practices are identified.

Additional information on improving OD is discussed separately [27].

3.3 Operational Discipline Training

A one-day OD Leadership Workshop has been developed with the following goals:

- Build value for why OD is important
- Introduce the "next generation" DuPont program
- Review methods for evaluating and improving OD
- Develop a path forward for improving OD at the site, based on site goals.

The Workshop is usually presented at one site, but it has also been offered to multi-site groups with few changes. Participants have generally included a cross-section of site roles and levels to provide diverse views for the development of a path forward for improving OD at the site.

Pre-work for the workshop includes developing a short summary of current OD performance based on site SHE metrics. Participants also complete the OD Self-assessment survey prior to the workshop, and a summary of the results is used in the workshop to provide additional data on current OD performance and to provide suggestions for improvement. Typically, 2-3 areas for improvement have been identified and an initial path forward for each has been developed. Sites have then taken this information back to site leadership groups for further refinement and prioritization versus other site SHE activities.

The recommended path forward following the workshop is to provide a shorter version of the training to all site employees to build value for OD and awareness of the DuPont OD program. The site can then send the OD Self-assessment survey to all employees, if desired, for broad input on the current status of OD at the site. Follow-up meetings based on the OD Self-assessment and on the planned path forward should then be held with employees for additional discussion of OD performance and improvement at the site.

4. Summary

Well-designed SHE systems are only as effective as the day-to-day ability of everyone in an organization to rigorously follow procedures correctly and safely every time. Once effective SHE systems have been developed, therefore, the greatest opportunity for maintaining and improving SHE performance may result from emphasizing, evaluating, and improving OD. High levels of OD in an organization help ensure strong SHE performance and overall operating excellence. OD problems can lead to poor SHE performance and higher overall risk, possibly resulting in serious injuries and catastrophic incidents. A complicating factor is that every company and every site is different, often with somewhat different cultures and practices. A "one size fits all" approach to OD therefore is unlikely to be successful for all organizations or sites.

The OD program at DuPont has been revised and simplified to include both organizational and individual OD characteristics that can be evaluated and improved in a flexible way. Organizational OD is focused on ensuring the dedication and commitment of everyone in the organization to carry out each task, the right way, each time. Leadership Focus and Employee Involvement are fundamental to achieving success in organizational OD and in maintaining high levels of Practice Consistent With Procedures and Excellent Housekeeping as visible indicators of performance. Ultimately, organizations are successful through the actions of individuals, who work safely by doing their tasks, the right way, every time. Individual OD is focused on ensuring that individuals have the Knowledge, Commitment, and Awareness to succeed and be safe. Both organizational and individual OD are necessary to achieve high levels of OD and sustained excellent SHE performance. As Jim Collins writes in *Good to Great* [11]:

> Sustained great results depend upon building a culture of self-disciplined people who take disciplined action…

A comprehensive OD program, such as the one described in this paper, can be an important element in the transition from good SHE performance to great SHE performance.

References

1 Klein, James A., *Operational Discipline in the Workplace*, Process Safety Progress, Vol. 24, No. 4, p. 228-35, December, 2005

2 du Pont, E. I., letter to this father, 1808 (source: Hagley Museum & Library, Wilmington, DE)

3 Burk, Arthur F., *Achieving Excellence in Process Safety and Risk Management*, Managing Safety and Health in the Workplace, Chicago, June 13-14, 1994

4 Klein et al, *Two Hundred Years of PSM at DuPont*, Risk Management: The Path Forward, AICHE, 2005 (20th Annual CCPS International Conference, April 11-13, 2005)

5 Center for Chemical Process Safety, *Guidelines for Implementing Process Safety Management Systems*, American Institute of Chemical Engineers, 1993

6 Center for Chemical Process Safety, *Guidelines for Risk-Based Process Safety Management Systems*, American Institute of Chemical Engineers, 2007 (in press)

7 DuPont, SHE Standard S21A, *Process Safety Management*, 2003 (internal document)

8 Gil, Frederick, *Integrity Management: Learning From Past Major Industrial Incidents*, BP, Booklet 14, 2004

9 Buckingham, Marcus et al, *First, Break All the Rules*, Simon & Schuster, New York, 1999

10 Center for Chemical Process Safety, *Guidelines for Preventing Human Error in Process Safety*, American Institute of Chemical Engineers, 1994

11 Collins, Jim, *Good to Great*, HarperCollins, New York, 2001

12 IAEA, *Self-Assessment of Safety Culture in Nuclear Installations*, Tecdoc-1321, 2002

13 Kinlaw, Dennis C., *Coaching for Commitment*, 2nd Ed., Jossey-Bass Pfeiffer, San Francisco, 1999

14 Kletz, Trevor, *An Engineer's View of Human Error*, 3rd Ed., IChemE, Rugby, UK, 2001

15 Smith, Rolf, *The 7 Levels of Change*, Summit Publishing, Arlington, TX, 1997

16 Walter, Robert J., *Discovering Operational Discipline*, HRD Press, Amherst, MA, 2002

17 Geller, E. Scott, *Keys to Behavior-Based Safety*, Government Institutes, Rockville, MD, 2001

18 Geller, E. Scott, *Social Influence Principles*, Professional Safety, p. 25-31, Oct., 2002

19 Geller, E. Scott et al, *Behavior-Based Safety Coaching*, Professional Safety, p. 42-9, Jul., 2002

20 Gregory, Earl D., *Building An Environment That Promotes Safe Behavior*, Professional Safety, p. 20-7, Oct., 1996

21 Hidley, John H., *Critical Success Factors for Behavior-Based Safety*, Professional Safety, p. 30-4, Jul., 1998

22 Mullen, Jane, *Investigating Factors That Influence Individual Safety Behavior at Work*, J. Safety Research, Vol. 35, p. 275-285, 2004

23 Petersen, Dan, *Leadership & Safety Excellence*, Professional Safety, p. 28-32, Oct., 2004

24 Reynolds, Stephen H., *ABCs of Behavioral Safety*, Professional Safety, p. 23-5, Feb., 1998

25 Topf, Michael D., *Change in Attitude Fosters Responsibility for Safety*, p. 24-7, Dec., 1995

26 Williams, Joshua H., *People-Based Safety*, Professional Safety, p. 32-6, Feb., 2003

27 Klein, James A. and Bruce K. Vaughen, *Evaluating and Improving Operational Discipline*, Plant Process Safety Symposium, Apr. 22-26, 2007

28 www.surveymonkey.com

OECD Guidance on Safety Performance Indicators

Guidance for Industry, Public Authorities, and Communities for developing SPI Programmes related to Chemical Accident Prevention, Preparedness, and Response

Kim Jennings, Chair, OECD Working Group on Chemical Accidents

This paper will present the culmination of more than five years of work by experts from North America, Europe and Asia who were brought together by the OECD to develop cutting edge guidance on the development and use of Safety Performance Indicators (or SPIs) in the context of chemical accident prevention, preparedness and response. This Guidance is scheduled for publication later this year, and will be available from the OECD website (www.oecd.org/ehs) and in hardcopy on request. An interactive web-based version is planned for mid-2008.

This guidance was developed to help enterprises, authorities and communities better assess the actions taken to control risks of chemical accidents and to provide a basis for understanding whether the risks are under control. SPIs can provide an early warning that critical safety procedures and systems have deteriorated, or are not operating as intended. Evidence has shown that use of SPIs can also help companies operate more efficiently.

Following some background information on the OECD and its work related to chemical accidents, this paper describes the process of developing the SPI guidance which included a two-year pilot program designed to help make the guidance as useful and practical as possible. It also sets out an overview of the document, which consists of two main parts: first, a step-by-step guide for developing and implementing an SPI program, and second, lists of possible activities and outcome indicators, set out by subject matter, for the readers to choose or adapt in light of their particular circumstances.

I would like to note that the OECD will be seeking help from the CCPS and its members (as well as from other organizations who work with industry, public authorities and communities). Specifically:

- organizations such as the CCPS will be asked to facilitate the distribution of the guidance to its members;
- the CCPS can help enterprises with the implementation of safety performance indicators by developing supporting materials specific to its members and by setting up mechanisms for sharing experience among members;
- the CCPS can send provide comments on the Guidance. OECD appreciates any feedback, which will be taken into account in future revisions of the document.

Background

The Organisation for Economic Co-operation and Development or OECD is an intergovernmental organization in which representatives of 30 industrialized countries (from Europe, North America, and the Pacific) and the European Commission meet to co-ordinate and harmonize policies, discuss issues of mutual interest, and work together to respond to international concerns. Most of the OECD's work is carried out by more than 200 specialized committee and subsidiary groups made up of member country delegates and invited experts from industry, labor and other stakeholders(?). It is one of these subsidiary groups, the Working Group on Chemical Accidents, which leads the efforts related to chemical accident prevention, preparedness, and response.

The OECD efforts in this area started over 20 years ago during a meeting of the OECD Environment Committee when member countries declared that they would "ensure the existence of appropriate measures to control the potentially hazardous installations, including measures to prevent accidents." This declaration led to a High Level Meeting of the OECD Chemicals Group to discuss how the OECD could assist in addressing this problem of accidents involving hazardous substances. The Group agreed that the OECD is uniquely qualified to provide an effective forum for intensifying national and international efforts to prevent chemical accidents.

In 1988, the OECD organized a Conference on Accidents Involving Hazardous Substances. The Conference was attended by ministers and senior government officials from OECD member countries with the aim of strengthening national and international policies with regard to accidents involving hazardous substances and the planning of emergency response. Conference members reviewed existing measures for improving the safety of hazardous installations and discussed the roles and responsibilities of government, industry, and labor. Participants concluded that the policies and programs of OECD Member countries with regard to accidents should be directed toward three basic goals: 1- to limit the frequency and severity of accidents through better measures to prevent releases of hazardous substances; 2- to prevent adverse effects from accidents through better land-use planning; and 3- to mitigate the consequences of accident though development of adequate emergency plans and measures. It was from this landmark meeting that an ambitious OECD program in this field was initiated and work on developing *Guiding Principles for Chemical Accident Prevention, Preparedness and Response* was begun.

This OECD program has resulted in the adoption of four Council Acts which have influenced and helped shape laws, regulations, and polices concerning chemical accidents in OECD member countries as well as around the world.

More widely known are the OECD *Guiding Principles for Chemical Accident Prevention, Preparedness and Response* which was first published in 1992. It is a comprehensive text addressing the roles and responsibilities of all relevant stakeholders (including industry, public authorities, communities, and others). It reflected the outcomes of a series of Workshops on a range of issues associated with chemical accident prevention, preparedness, and response involving key experts from around the world with varied perspectives and expertise. Following 10 years of implementation of these *Guiding Principles* and armed with the conclusions and recommendations from fifteen additional Workshops, the OECD published the 2nd Edition of the *Guiding Principles* in 2003. The *Guiding Principles* have been translated into a number of languages and has been distributed widely throughout the world.

It was during the early stages of revising the *Guiding Principles* that the Working Group on Chemical Accidents initiated the project to develop a companion document - the *Guidance on Safety Performance Indicators*. The objective of this project being to facilitate implementation of the *Guiding Principles*, and to assist industrial enterprises, public authorities, and communities near hazardous installations assess whether actions taken to implement the *Guiding Principles* lead to improved levels of safety over time.

Developing the Safety Performance Indicators Guidance

The Working Group established a number of parameters for this project, stating that the work should:

- focus on developing guidance for safety performance indicators, rather that precise indicators, to allow flexibility in application (*i.e.*, the guidance can be adapted to different countries, industries, and cultures);

- address both measures of activities/organization of work, and measures of outcome/impact;

- build on existing materials and experience, to the extent possible;.

- address the activities of public authorities and industry (and, to the extent applicable, other stakeholders);

- be based on the *Guiding Principles*, although not be as detailed.

The Working Group established a group of experts from OECD member countries, including representatives from governments, industrial companies, labor, and other non-governmental organizations. The Group started by circulating a questionnaire to gather information on any experiences related to developing and measuring performance indicators. The results showed that there were only limited efforts to measure and evaluate the success of chemical safety programs. At that time very little was being done by public authorities and communities and industry was focusing primarily on audits of their facilities' processes and systems. The idea of measuring the long-term success of a program related to chemical safety was in its infancy.

In order to meet the objective of helping industry, public authorities, and communities measure whether the many steps taken to reduce the likelihood of accidents, and improve preparedness and response capabilities, truly lead to safer communities and less risk to human health and the environment, the Expert Group designed the document to:

- Serve as a tool to be used by stakeholders when establishing programs to:
  - Determine their success in developing and implementing appropriate requirements, policies, and procedures designed to reduce the likelihood of accidents and improve preparedness and response capabilities; and
  - Assess whether actions taken to implement those requirements continue to operate as intended, and are not deteriorating over time

- Not define a precise methodology, rather provide guidance and suggestions for elements to help stakeholders define their own safety performance indicator system.

- Be flexible, so they it can be helpful to the range of interested industries, public authorities, and communities worldwide. It is intended to complement other related activities.

Using these parameters, the Expert Group concluded that the document would consist of an introductory section including information on how to use the guidance, as well as the main body of the document, which would consist of three distinct sections, one each for industry, public authorities, and communities. These sections would be divided by subject areas based generally on the *Guiding Principles* and each section would contain a general overview of the subject area, a target for that topic, and guidance for outcome and activity indicators. The *Guidance* would include specific outcome and activity indicators which users could choose, and adapt, to fit the specific needs of their organization.

The industry section is designed to be used by any enterprise, irrespective of size, location, or whether privately or publicly owned. The common denominator for all members of the target audience is that they handle hazardous chemicals so as to create a risk of leak, fire, or explosion with potential for adverse effects on health, the environment or property. Therefore, this would include any enterprise that produces, processes, uses, stores, or disposes of such chemicals.

The first edition of the OECD *Guidance on Safety Performance Indicators* was published in 2003, which was followed by, an interactive, web-based version. In addition to providing the contents of the SPI document which can be queried by subject area and stakeholder group, it also contained an additional tool called "Create a Customized SPI Program". This tool helps users adapt the general guidance to their own particular circumstances. .

This first draft was designated as an "interim document", in order to test the guidance, get feedback from users, and revised the *Guidance* to make it as practical as possible. A pilot program was initiated to help ensure that sufficient, high quality feedback was gathered from each of the three main target groups (industrial enterprises, public authorities, and communities). From 2003 to 2004, volunteers from eight countries tested the guidance document by developing and implementing safety performance indicator systems for their organizations and providing specific feedback via questionnaires.

The participants in the pilot program provided very constructive comments which were consolidated and used to further refine the document. For the most part, the participants concluded that the *Guidance* was well organized and contained sufficient instructions and comprehensive lists of potential outcome and activities indicators. Nevertheless, there were numerous suggestions for improving the *Guidance*, including:

- Expand and/or improve the "How to use this *Guidance*" section, because many organizations had difficulty getting started developing a safety performance indicator system;

- Reviewing the introductory materials to clarify certain parts and include further information on the benefits of the SPI program and on what to measure;

- Adding further examples, including more guidance concerning the use of metrics; and

- Providing more specific guidance for police, firefighters, and other emergency personnel.

During the timeframe that the OECD was piloting the *SPI Guidance*, the Health and Safety Executive (HSE) in the United Kingdom was working with the Chemical Industries Association and a number of enterprises to develop similar guidance. In 2006, the HSE published the document, "Developing process safety indicators: A step-by-step guide for chemical and major hazard industries". This guidance document along with the results of the OECD pilot program formed the basis for the revisions to the OECD *SPI Guidance*.

Overview of the SPI Approach

Following the Expert Group review of comments as well as consideration of efforts that had taken place in this area since the initial publication of the document, it was agreed that there should be significant revisions to the format and that further guidance on the process of establishing an SPI program and on metrics should be included.

Specifically, the guidance will be divided into two separate publications, one for industry and the other for public authorities and communities. The public authorities section also includes sections for first responders and elected officials.

The final *Guidance* will contain three chapters:

- Objective and scope (including a description of the target audiences);
- How to develop an SPI program – step by step guidance

- Choosing targets and indicators – list of specific indicators which organization can choose and adapt to their circumstances.

In addition, the *Guidance* contains several annexes providing: guidance on developing metrics; examples; summary of targets; explanation of terms; and references.

The second chapter - How to develop an SPI program – is the most significant addition to the document. It sets out a process in seven steps to assist user in establishing and implementing a safety performance indicator system for their organizations. The seven steps include (see figure 1):

- **Step One: Establish the SPI Team**
  The first step is to identify a champion or steward to lead the initial effort, promote and co-ordinate the introduction of the SPI Program, ensure effective communication, and generally oversee the Program's implementation. This could consist of a single person or team of people, depending on the size of the organization and availability of resources. It emphasizes the need to have management on board, and to report to individuals with the authority to make changes in light of the results of the SPI program.

- **Step Two: Identify the Key Issues of Concern**
  Once the SPI team and other arrangements are in place, the next step is to identify the scope of the SPIs to be measured by identifying the safety programs, procedures, and policies that are most important for the protection of human health, the environment, and/or property.

- **Step Three: Define Outcome Indicator(s) and Related Metric(s)**
  At this point the SPI team would define outcome indicator(s) addressing the key issues of concern. Outcome indicators are designed to help measure whether safety-related actions are, in fact leading to less likelihood of an accident occurring and/or less adverse impact on human health or the environment from an accident. Outcome indicators tell you *whether* you have achieved a desired result but, unlike activities indicators, do not tell you *why* the result was achieved or *why* it was not. These indicators need to convey clear information on safety performance to those with the responsibility and authority to take action. An SPI consists of two key components:

  - A definition, which should clearly state what is being measured in terms that are meaningful to the intended audience;
  - A metric, which defines the unit of measurement. This should be precise enough to highlight trends in safety over time and/or highlight deviations from safety expectations that require action.

- **Step Four: Define Activities Indicator(s) and Related Metric(s)**
  The next step is to define activities indicators to monitor the key elements of the safety programs, procedures, and policies identified in Step Two. Activities

indicators relate to outcome indicators and help to measure whether critical elements of safety programs, procedures, and policies are in place to achieve the desired outcomes. Whereas outcome indicators are designed to provide answers about whether you have achieved a safety outcome, activities indicators are designed to provide information about why or why not the outcome was achieved. Therefore, well-designed activities indicators should provide the information needed to correct programs, procedures, and policies when the desired outcome is not being achieved.

- **Step Five: Collect the Data and Report Indicator Results**
  The next step is to decide how to collect the data and report the safety performance results. Data collection approaches (*i.e.*, data sources, how the data will be compiled, how often, and what the reports will look like), as well as roles and responsibilities for collection and reporting, should be specified.

- **Step Six: Act on Findings from Safety Performance Indicators**
  Information from SPIs (such as tolerances being exceeded, disturbing trends over time, inconsistent results) must be acted upon; otherwise, there is little point in implementing an SPI Program. Relevant personnel should receive SPI results in a timely way and should follow up adverse findings to rectify defects in the safety programs, procedures, and policies.

- **Step Seven: Evaluate and Refine Safety Performance Indicators**
  The SPI Program, including the indicators and metrics, should be periodically reviewed and evaluated. Developing an effective SPI Program is an iterative process, and the Program should be refined as experience is gained or new safety issues of concern are identified (*e.g.,* due to new laws or changes in technologies). Furthermore, enterprises might decide to expand their SPI program once experience is gained.

The third chapter of the *Guidance* helps users to choose and adapt outcome and activities indicators, along with associated targets, by providing a menu of possibilities based in large part on the *Guiding Principles for Chemical Accident Prevention, Preparedness and Response*. The lists of indicators remain similar to the section in the previous edition of the *Guidance*.

The chapter is divided by the range of subject matter involved in chemical accident prevention, preparedness and response, from policies and administrative procedures, to technical issues, co-operation and communication, preparedness and response, and accident reporting and investigation. (See Attachment 1)

<u>Where do we go from here</u>

The *Guidance on Safety Performance Indicators* is in the final stages of being completed and will be sent out for review to the OECD Working Group on Chemical Accidents and key stakeholders in each of the OECD member countries in mid-2007. It is anticipated that the document will be published by the end of this year. The companion website will be update based on this final version and should be available by mid-2008.

As noted above, the OECD will be calling on the CCPS and other organizations to help with the distribution and implementation of the *Guidance*, and to provide feedback.

Figure 1
Seven steps to create and implement an SPI Program

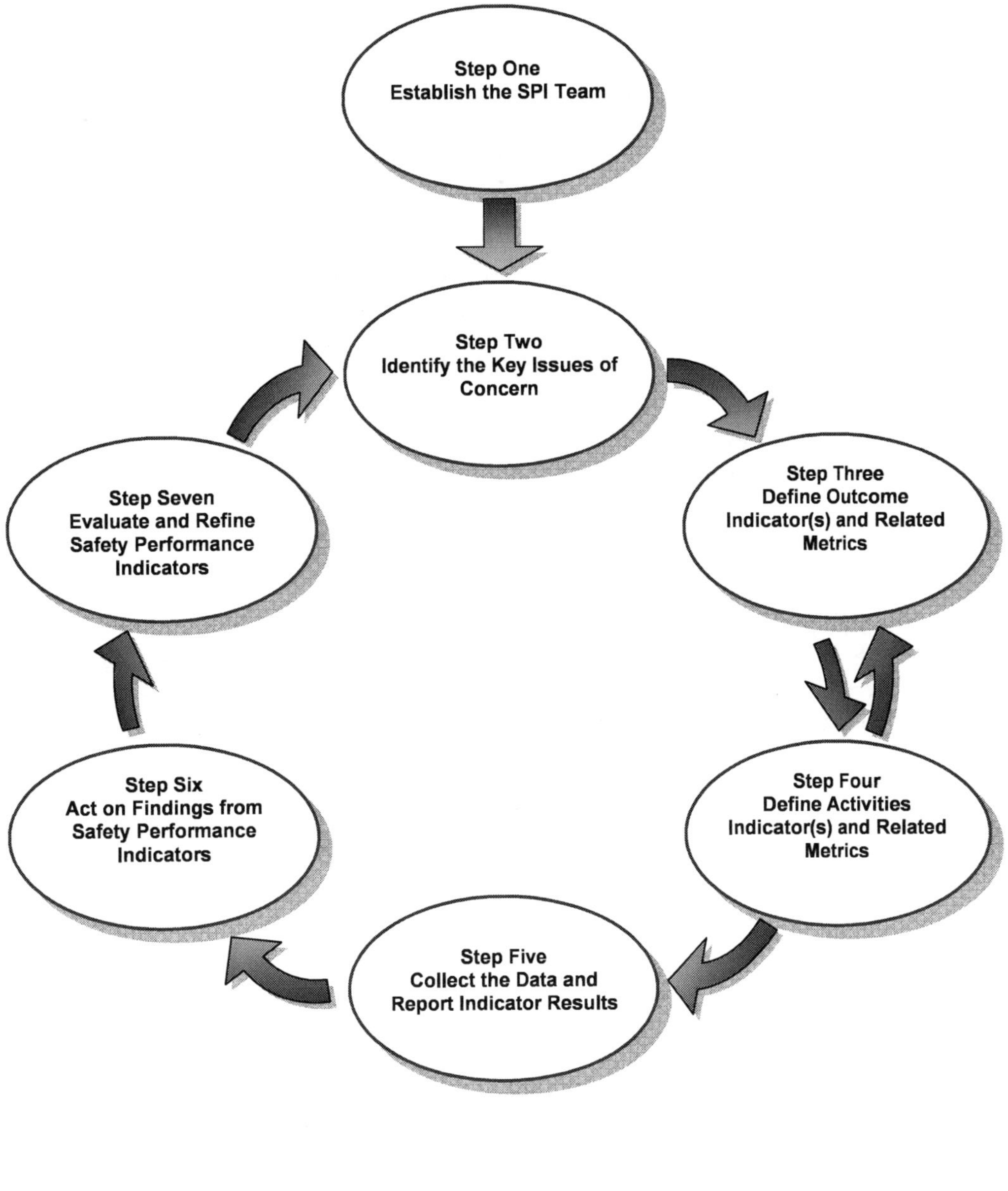

Attachment 1
Example of Subjects in the Industry section

Policies and General Management of Safety

- Overall policies
- Safety goals and objectives
- Safety leadership
- Safety management systems
- Personnel
  - *Management of human resources*
  - *Training and education*
  - *Internal communication/information*
  - *Working environment*
- Safety performance review and evaluation

Administrative Procedures

- Hazard identification and risk assessment
- Documentation
- Procedures (including work permit systems)
- Management of change
- Contractor safety
- Product stewardship

Technical Issues

- Research and development
- Design and engineering
- Inherently safer processes
- Industry standards
- Storage of hazardous substances (special considerations)
- Maintaining integrity/maintenance

External Co-operation

- Co-operation with public authorities
- Co-operation with the public and other stakeholders (including academia)
- Co-operation with other enterprises

Emergency Preparedness and Response

- Internal (on-site) preparedness planning
- Facilitating external (off-site) preparedness planning
- Co-ordination (within industrial enterprises)

Section F. Accident/Near-Miss Reporting and Investigation

- Reporting of accidents, near-misses and other "learning experiences"
- Investigations
- Follow-up (including application of lessons learned and sharing of information)

Management Review: The Way to Be Sure Your Process Safety Systems Are Working

Jack L. McCavit
JL McCavit Consulting, LLC
Pampa, Texas
jlmconsulting@sbcglobal.net

Donald K. Lorenzo, P.E.
ABS Consulting
Knoxville, Tennessee
dlorenzo@absconsulting.com

ABSTRACT

An effective management review system can produce a major improvement in process safety results. As plant resources decline, the risk of a process safety system failure increases, and the breakdown can occur far more quickly than a 3-year audit cycle can detect. Thus, there is a need for periodic management reviews, yet most companies have only a limited management review system or none at all.

The Center for Chemical Process Safety created ***Risk Based Process Safety*** *as the framework for the next generation of process safety management. Management review is a new element in that framework. This paper describes the benefits of a management review system and describes how to develop and implement an effective management review system.*

The paper includes case histories from the authors' experience to illustrate ways to make your management review system a success.

1. INTRODUCTION

In 2003 a company conducted a process safety audit at one of its facilities that had been ranked well above average for several years. The audit results were shocking. In less than 3 years, the site's process safety management (PSM) program had deteriorated to performance levels typical of 1993. Several elements had fallen into complete disarray. How could the site's PSM elements decline so quickly, especially considering its well-established PSM program and history of superior results?

An investigation revealed that the management team had been preoccupied with a reorganization of the site over the past year. The PSM coordinator position had been eliminated and the duties were spread among others. Simultaneously, there was a flurry of retirements and personnel changes. The management team at this facility sent a strong message that process safety systems were not important and the plant got the message. In a few short months, a decade of accomplishment was in danger of imminent collapse. Fortunately, the deterioration was found in an audit and not in an incident investigation. Nevertheless, this frightening near miss forced upper management to recognize the potential gap between periodic audit findings and real-time PSM performance. To address this gap, management implemented a system of management reviews to regularly monitor the health of its PSM systems.

Experiences like this across the chemical industry prompted the Center for Chemical Process Safety (CCPS) to include *management review* as one of the new elements in its risk-based process safety framework for the next generation of process safety management.

2. WHAT IS A PROCESS SAFETY MANAGEMENT REVIEW?

A process safety management review system is a formal system intended to assure senior management that process safety systems are effective and that the results of the process safety systems are meeting expectations. The CCPS book, *Guidelines for Risk Based Process Safety* (Ref. 1), states "The *management review* process provides regular checkups on the health of process safety management systems in order to identify and correct any current or incipient deficiencies before they might be revealed by an audit or incident."

A process safety management review is a meeting during which site management reviews the status of selected PSM elements with subject matter experts (SMEs). Site management questions the experts and evaluates their responses. In many cases, the experts simply present metric data they routinely accumulate. For example, a review of the mechanical integrity system would almost certainly inquire about the percentage of scheduled inspection, test, and preventive maintenance tasks that were overdue. If they find the desired target of "zero" is being achieved, management should give appropriate accolades. If not, it should establish clear expectations, help develop corrective action plans, and provide the resources necessary to achieve the expectations.

However, the goal of the management review is to understand the robustness of the PSM system, not to simply elicit the desired answer. In the previous example, the fact that there are no overdue inspections sounds good, but how was that goal achieved? Have some inspections simply been rescheduled so they do not appear as overdue in the report? Why were they rescheduled? Who approved them? Were appropriate management of change (MOC) protocols followed? Only when those questions are satisfactorily answered can management conclude that the mechanical integrity system is effectively accomplishing its work commitments.

Thus, the management review team will often pursue meaningful data that are not simple metrics. For example, are MOCs being written for mechanical changes? The SME will have to examine work orders to find several examples that should have MOCs. Then the expert will need to find and review the MOCs to ensure they were appropriately performed and documented. If MOCs are missing or inadequate, the management review team will need to evaluate the circumstances to determine if the issue is isolated to one area of the plant or one type of change so effective corrective actions can be implemented.

The results of the management review meetings should be documented. The documentation is typically more in the style of a memorandum than a formal report, but it should, at a minimum, include (1) the agenda, (2) the data that were presented, and (3) the action items resulting from the discussions. It is important that the general population of the plant know that the management team is personally committed to and involved in process safety, so the minutes from the process safety management review meetings should be widely distributed.

3. WHAT ARE THE BENEFITS OF MANAGEMENT REVIEWS?

Intellectually, most people understand that driving an automobile is one of the highest risk activities we undertake. In fact, chemical and refinery workers in the United States are much more likely to die in a car wreck than in an accident at their work place. Most people also understand that the risk of dying in a car wreck increases significantly as speed increases. To manage the risk of automobile wrecks, society imposes limits on the speeds that we can legally drive. Nevertheless, even with knowledge of the risks and clearly defined legal limits, some people exceed safe speeds. For most of us, the concern about speeding is a potential ticket, not a potential fatal accident. We even buy radar detectors so that we can drive faster than the speed limit and not get caught by the police. Even careful drivers will sometimes speed when they are in a hurry.

If drivers in our society will not do what they can to reduce the risk in what is clearly the highest risk activity of their day, why would we expect people in chemical plants and refineries to act differently when they get to work? Why would we expect engineers to give priority to work on an incident investigation finding when much of their recognition/reward comes from cost reduction? Why would we expect operations personnel to obtain all of the signatures on an MOC before commissioning new equipment when they are confident everything is okay and their bonus depends on production? Why would we expect engineers to finish a process hazards analysis (PHA) prior to a regulatory deadline when the meetings are tedious and they found no significant issues in the previous review? Why would we expect operators to follow procedures as written when they know easier ways to complete their work and have been praised for their efficiency?

The list is endless. The point is that conformance to management systems should not be expected to happen naturally. Like police enforcement of speed limits, the frequency and depth of management reviews indicate how seriously management intends to enforce the process safety elements. Management review gives stakeholders confidence that the process safety elements are working effectively.

3.1 Management Review Gives Management an Opportunity to Set Process Safety Expectations

Management review gives senior management an opportunity to establish expectations that can impact the plant for years. In the early 1990s a facility had as much as 15% of its equipment past due for inspection. Over a couple of years, the facility reduced the backlog of overdue inspections to less than 0.5%. The management team wanted to celebrate that improvement. However, in a management review meeting, the plant manager asked one simple question that set the tone for years to come. He asked why they should feel good about having any overdue inspections. The plant manager acknowledged the improvement, but he clearly communicated that continued reduction was expected. His single question effectively lowered the target to zero, which was achieved a few months later. Without the plant manager's question, the percentage of past-due inspections would likely have hovered between 0.5 and 1.0% indefinitely.

It is also possible to sabotage a PSM element during its management review. For example, if the SME reports that there is no one available to lead a PHA and it will not be completed by a regulatory deadline, the site management team must respond and help find the resources. If the management team knowingly allows a PHA to be late, it is sending a message that process safety is not important, nor is it important to be in compliance with the law. When the reviews identify failed systems, the management team must respond quickly or the problem switches from the failed system to the management team itself.

Conversely, when the management team learns of failed systems and takes action to fix them, the message to the plant is also clear. Process safety is important, and the plant management team will not tolerate poor performance in process safety elements.

3.2 Management Review Gives Management an Opportunity to Be Involved in Process Safety

It is impossible for upper management to be personally involved with all elements of process safety on a daily basis. However, management review meetings provide a regular opportunity for personal involvement in many elements of process safety that would otherwise be difficult to engage. In addition to ascertaining the health of the PSM system, the manager can praise those who are executing their jobs well. The mere show of interest by upper management is a powerful motivator for the rest of the plant staff. The plant staff's efforts are important, and their contributions are recognized.

3.3 Management Review Reduces Surprises During Audits

Effective process safety management reviews provide frequent feedback on the status of PSM systems, so deficiencies and incipient failures can be corrected. However, the management reviews will not be as thorough as an audit so there will continue to be occasional audit findings. Management review emphasizes effectiveness rather than compliance. Nevertheless, if the management review system is designed properly, there should be no consequential surprises during the audits.

3.4 Management Review Results in Fewer Incidents

The ultimate goal of a management review system is not improved PSM systems; the goal is fewer process safety incidents. Management review will help establish a culture in which following the rules is expected and is the norm. PSM elements will be robust and the culture will not tolerate poor performance. The inevitable outcome of a good process safety management review system will be fewer, and less severe, process incidents.

4. WHAT DOES A MANAGEMENT REVIEW SYSTEM INCLUDE?

The answer is, in a word, "everything" encompassed by the PSM element being reviewed. Management is free to inquire about anything that it believes is indicative of the performance of the element, and varying the focus of the questions is highly beneficial. The objective is to gain confidence that the system is robust and effective.

In particular, management must first seek to understand whether the system being reviewed is producing the desired results. For example, consider the MOC system. If the organization's quality of MOC reviews is less than satisfactory as evidenced by process incidents precipitated by changes, or it is not improving as a result of management system changes, then management should identify possible corrective actions and pursue them. Possibly, the organization is not working on the correct activities, or the organization is not doing the necessary activities well. Even if the results are satisfactory, management reviews can help determine if resources are being used wisely – are there tasks that could be done more efficiently or tasks that should not be done at all?

Management can combine metrics and indicators with personal observations, direct questioning, audit results, and feedback on various topics to help answer these questions. Activities and topics for discussion could include the following:

- Review the completeness and quality of a sample of MOC reviews for each operating area. Discuss any significant gaps
- Talk to people in each operating area and find out if the MOC process is being used as intended. Is there evidence of circumvention?
- Review recent audit findings that addressed MOC and determine the status of all corrective actions. Are any items overdue or backlogged?
- Determine if any incidents occurred or if any trends were found that listed MOC failure as a root cause or contributing factor
- Ask employees their opinion of how effective the new employee and contractor training is regarding MOC. Query employees on their knowledge of MOC to monitor refresher training
- Review the staffing, resources, and time spent to address MOC issues. Examine any major changes from previous years

In addition, the MOC element owner should be able to explain any trends or anomalies in the current metrics. If any major projects to address known MOC gaps are ongoing, a briefing should be prepared for the management review committee.

The results of a management review of MOC activities should demonstrate that leadership at the facility is aware of and values MOC, and is intent on ensuring that all changes are properly evaluated prior to implementation. The management review process should also shine a bright light on efforts to perform, document, collect, and maintain MOC information and metrics, including improving the efficiency of work activities that support this element. In addition, an effective management review process educates the entire leadership team on the importance of MOC and the role it can play in helping to identify hazards, manage risk, and sustain the business.

There are several examples of metrics and items that could be included in management reviews of other elements in the *Guidelines for Risk Based Process Safety*.

5. HOW IS A MANAGEMENT REVIEW SYSTEM IMPLEMENTED?

The key to a successful management review system is to "just do it." It does not have to be perfect to start. It can be changed with time as new issues arise or as reviews appear to be unproductive. It must not be overly complicated or too burdensome to gather the information for the review meeting. Complexity, procrastination, and inflexibility are enemies of successful management review.

The basic steps are:

- Get management support
- Decide what elements will be reviewed
- Decide on a schedule for the reviews
- Develop a list of items to be included in the reviews for each element
- Decide which items will be mandatory and which items will be optional
- Update the system about a year after implementation and each year thereafter

The first step in establishing a management review system for a company (or for an individual site) is obtaining senior management's support. Senior management must believe in the benefits of the work and provide the resources to establish and maintain the program.

After receiving support from senior management, the implementation team needs to define what elements are going to be in the initial review program and at what frequency the reviews will be conducted for each element. Stakeholders from process safety leadership; SMEs; and line management from the operations, technical, maintenance, and engineering groups need to be involved. The long-term objective is to include every element at least once per year, but the initial program should focus on areas with a history of deficiencies and audit findings.

Within a company all the sites should schedule the same reviews during the same month unless there is some good reason to do otherwise. Consistency of timing helps senior management (above site level) review data from diverse sites and helps company process safety leadership understand general issues. However, one site may have a weak MOC program that warrants more frequent reviews while another site may have a complex contractor management program that warrants more frequent reviews. When multiple reviews are scheduled, try to schedule at least one of the reviews to coincide with the companywide review month. It is not critical that all sites do the same reviews at the same time, but the benefits are strong enough that consistency should be a goal.

The following is an example initial review schedule. Note that various elements are reviewed with different frequencies and that more than one element can be reviewed in a single month. Also note that the company chose to focus its initial efforts on elements with a history of issues; therefore, several elements were not included in the schedule.

- PHA, twice per year in June and December
- MOC, quarterly in January, April, July, and October
- Incident investigation, monthly
- Action items from PHAs, incident investigations, MOCs, and all other sources, monthly
- Standard operating procedures, annually in November
- Mechanical integrity for fabricated equipment, quarterly in February, May, August, and November
- Mechanical integrity for instrumentation, quarterly in March, June, September, and December
- Contractor management, annually in January

Once a schedule is set up, it is helpful to develop a list of activities/topics to guide the review of each element. Some items in the list should be mandatory for all of the sites, while others can be optional. Sites should be encouraged to add items to the list and adapt it to their particular situation. Some of the items may be routinely gathered metrics while others may require special analyses in preparation for each review. It is critical that line management, process safety personnel, and SMEs all be involved in identifying and selecting the review items so that all of these groups have ownership of the items. With a new management review system, it is wise to spend the bulk of the effort on elements that are known to have weaknesses. For those weaker elements, it is appropriate to have more frequent reviews and more mandatory review items.

The items selected for review are a key determinant of the effectiveness of the management review system. If the items selected are superficial, the reviews will not be meaningful. However, if the items selected impose burdensome data collection and analysis requirements, the reviews will be resented by the organization and the results may not be worth the effort. The key is to have knowledgeable people identify items that offer excellent insight into the effectiveness of the system without creating a lot of extra work. A good management review system will probe unique and revealing topics without consuming a significant amount of the organization's resources. The information that is gathered should provide a good indication of the status of the most important aspects of the element being reviewed.

The other key determinant is the composition of the review team. The team must include top management of the site and SMEs. Process safety activities are largely executed by the line organization, so it is critical that line management, including the facility manager, attend the meetings. Managers of functions, including operations, purchasing, projects, maintenance, and engineering, should be in attendance. The human resources manager should also be included if that person has responsibility for training. The key is to have all of the responsible managers present so that if there are issues that need resolution, the right people are in the room to take action.

Equally important attendees are SMEs. For example, if the meeting agenda includes PHAs, the PHA SME should gather the information and present the information to the management team. The SMEs will be more efficient in gathering data and will be most attuned to current issues within the management system. Obviously it is important that the management team not shoot the messenger, or the SMEs will be prone to minimize the concerns and maximize the positives. An open, trusting culture is essential to success.

The meetings themselves should be scheduled to last about 1½ hours to maintain focus and high energy. Two or three elements should be reviewed in that time. A management review of some element(s) should be scheduled every month. This will keep management engaged in process safety and emphasize the importance of process safety to the plant staff. The meeting should conclude with results that require actions within the facility.

Flexibility is important in maintaining a strong management review system. After about a year, the frequency of the element reviews should be reconsidered and the mandatory review items should be evaluated. It is likely that some of the review items are just not effective. For example, suppose that the number of MOCs written per month was a metric that was chosen to be reviewed. After a few reviews, it may become apparent that the value does not change much or that the changes are correlated to turnarounds. Thus, another metric may offer better insight into whether MOCs are being written as required. The metric might be replaced with a review of work orders to see if some changes are being inappropriately treated as replacements in kind. The goal is always to review meaningful data, and the flexibility to change is important.

Recognition that the system can and will change over time takes some pressure off of the initial development. If the developers believe the system has to be perfect at the beginning, the initial development will be unnecessarily long and difficult.

6. CONCLUSION

At an industry consortium meeting in 2002, eight different significant incidents were reviewed by company representatives. Each of the company representatives started their presentation by professing their surprise that their company made the kind of mistakes that caused the incident they were about to discuss. The representatives were embarrassed and discouraged by the failure of their management systems. Management review is intended to eliminate incidents that are the result of situations in which the site staff knows what to do, but they just are not doing it.

Management review is more important than ever in today's environment. Pressure for increased productivity has resulted in fewer process safety and other technical resources at many chemical and refining facilities. With the reduction in headcount in both process safety and technical resources, process safety systems are more likely to fall into disrepair. People are pulled in even more directions than in the past, and a natural consequence can be the deterioration of process safety systems. Systems must be regularly checked by management to prevent or discover the deterioration. Management reviews provide a unique opportunity for site management to send strong messages to the staff about the importance of process safety systems.

The CCPS included management review as one of the new elements in its framework for risk-based process safety because of its crucial importance in ensuring the reliability of PSM systems. A good process safety management review system will result in fewer, and less severe, process incidents.

7. REFERENCE

1. *Risk Based Process Safety*, New York: American Institute of Chemical Engineers, Center for Chemical Process Safety, 2007.

Improve the Risk Ranking Process by Categorizing and Detailing Consequence and Probability Categories

Adrian L. Sepeda, P.E.
A. L. Sepeda Consulting Inc.
www.sepcon.com

Concept: Improve accuracy by breaking a problem down into smaller subgroups, estimate each of those and then aggregate them to form the final estimate
Technology: "Divide and Conquer—A Fundamental Idea (an algorithm approach)" by Philippe Dumas

ABSTRACT

The Risk Matrix has been accepted as a valid methodology for converting hazards to risks and then comparing those risks to each other as a relative risk ranking process. Unfortunately, this is usually a very subjective process relying upon the personal expertise and experience of those performing the review to estimate both probability and consequence of each hazard potentially manifesting itself as an unwanted event. In the end, this approach is not consistently "repeatable" and/or "defendable." That is, it does not ensure that another group of people with essentially the same information would reach substantially the same conclusion—i.e. "repeatable." And because it is so subjective, it does not have sufficient technological grounding to allow it to be defended on its technical merits—i.e. "defendable."

This paper proposes ways to enhance "repeatability" and "defendability" by using a "divide and conquer" approach. It proposes categories for each level of the probability and consequence axes. This takes advantage of the fact that some information is known and other information is not known. Making several smaller estimates (some with known information) allows a better total estimate to be made. To demonstrate the concept, a 5 X 5 matrix example is used that categorizes each consequence level into four distinct types of consequences and provides three potential approaches for estimating probability, depending on the information available. These proposed categories and criteria for each category are shown for illustrative purposes only. Each user company should develop their own categories and details.

As a look toward future developments, the potential advantages of a three dimensional risk ranking tool are also briefly discussed, with potential axes suggested.

Improve the Risk Ranking Process
by
Categorizing and Detailing Consequence and Probability Categories

Adrian L. Sepeda, P.E.
A. L. Sepeda Consulting Inc.

BACKGROUND

Risk Ranking involves identification of hazards, conversion of those hazards to a risk by estimating consequences and probabilities (typically of the consequences of concern), comparison of one estimated risk to another for prioritization purposes, and development of recommendations to manage each risk. By comparing one estimated risk to another and the costs and benefits of risk management, one can justify and document the thought process and reasons for addressing the identified risks in a particular order.

The typical risk equation is written: Risk ~ Probability + Consequence, where "~" means "in some way equal to" and "+" means "plus consideration of." So the equation would read: Risk is in some way equal to the probability plus the consequence.

Initially, very simple risk matrices that used probability and consequence as axes were used to convert hazards to risks. In its simplest form a 3 X 3 matrix (three levels of probabilities and three levels of consequences) was used with descriptors such as "Low,"

"Medium," and "High." Figure 1 is an example of this simple Risk Matrix.
Notice that this simple 3 X 3 matrix can reasonably yield between three and up to five levels of risk as denoted on the figure as the "A," "B," "C," "D," and "E" squares. Each of the estimated risk levels, "A" through "E" should have a defined and specific action

required to manage the estimated risk. (See Attachment A for an example of what responses might be for estimated risk levels.)

For this specific 3 X 3 risk matrix, the estimated risk ranking equation would be written:

Risk Rank ~ Probability (Pick one: Low, Medium, High)
+ Consequence (Pick one: Low, Medium, High)

Therefore the relative Risk would be one of five values, "A" through "E," dependent on which level of probability and consequence were chosen. As an example, if the probability was "Medium" and the consequence was "High," the relative risk would be "D" which is higher than "A," "B," and "C" risks but not as high as "E" risks. Note that this process does not quantify the risk but rather ranks it relative to other risks evaluated. Documentation of how the relative risk rank of "D" was determined would consist of only a listing of the probability and consequence used to arrive at it and freeform notes, if any, on how they were established.

It soon became apparent that this would not meet the test of "repeatability" and "defendability" because of the inconsistencies in interpretations of "Low," "Medium," and "High" and documentation shortfalls. Each group converting hazards to risks using this simplified matrix with generic titles for the levels and no subcategories with additional details would/could arrive at different risk levels simply because of their personal experiences and how they felt that particular day. In some cases, the same group would arrive at different risks on different days.

The next generation of the risk matrix expanded the number of levels of consequences and probabilities believing that more must be better. Figure 2 is an example of a matrix with five levels of probabilities and five levels of consequences.

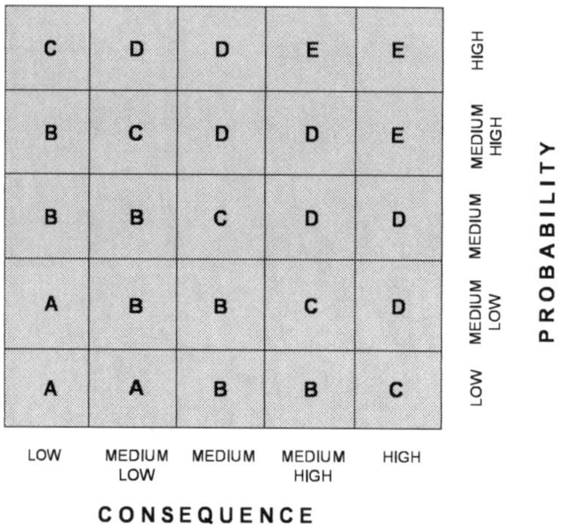

Figure 2

For this risk ranking matrix, the risk ranking equation would be written as:

Risk Rank ~ Probability (Pick one: Low, MediumLow, Medium, MediumHigh, High)
+ Consequence (Pick one: Low, MediumLow, Medium, MediumHigh, High)

Although this was a step in the right direction by breaking the estimates into smaller pieces, it did not include the categorical breakdown and specificity within each category necessary to assure "repeatability" and "defendability." It really does not matter how many pieces Probability and Consequence are broken down into if there are no specific criteria defined. Different groups could still reach different conclusions with essentially the same information just because of their backgrounds and personal biases. Documentation of the decision process for larger matrices had the same shortfalls as smaller matrices. Over time, the risk matrix took a variety of shapes and sizes but none of the approaches addressed the root problem—clean clear definitions for each level within each category.

PATH FORWARD

The algorithmic concept of Divide and Conquer can be used to address both the "repeatability" and "defendability" requirements. The user, by choosing what for them is:

- an appropriate number of levels (the size of the matrix),
- categories (areas of concentration) for consequence and probability, and
- an appropriate number of defining details within each of those categories

can then

- risk rank the hazards with repeatability,
- document the decision process with enough detail to allow someone in the future to retrace the process and understand how the decision was made, and
- demonstrate enough technical development of the process so as to make it both "repeatable" and "defendable."

This process allows the user to make estimates of small sections with specific criteria instead of trying to estimate the consequence of a major hazard with a single "guess." When delineated properly, some information in the smaller sections with specific criteria is actually known and the "estimate" becomes an accurate assessment. By combining these smaller estimates and accurate assessments, a better composite of the consequence or probability is possible. To reach the goal of "repeatability" and "defendability," each category of the components of risk (consequence and probability) must be delineated such that different groups of people with essentially the same information would reach essentially the same conclusion—i.e. risk rank. This "trail of logic" must be documented so that the path to the conclusion is understood and retraceable by others at a different time.

Using the Divide and Conquer approach, consider, for example purposes only, a 5 X 5 matrix with four categories for consequences with details within those categories, and three categories for probability estimation with details within those categories <u>at each level</u> within the matrix. See Figure 3.

PROBABILITY					
5	C	D	D	E	E
4	B	C	D	D	E
3	B	B	C	D	D
2	A	B	B	C	D
1	A	A	B	B	C
	1	2	3	4	5
	CONSEQUENCE				

Figure 3

Note that both the consequence and the probability axes are divided into five levels, 1 (the lowest) through 5 (the highest). Each level has specific definable criteria.

Overviews of the four **consequence** categories within each level are:

- Injuries and illnesses—onsite and offsite, real time and short term
- Environmental issues—onsite and offsite with various remediation requirements and times
- Community relations—public or regulatory reactions, local, state, or national
- Financial impacts—onsite and offsite property damage, loss of production capability, restoration costs, and potential fines

An example of what the supporting detail for each of the categories might be for a **Level 3 consequence** in this risk matrix is as follows. This example is extracted from the full range of five levels of consequences (Level 1 through Level 5) shown in Attachment A.

3	**Injuries or Illnesses:** **II-3.1= One onsite permanent incapacitating injury and/or illness** **II-3.2= Potential for onsite dermal or inhalation sensitization, or reversible toxicity that can impair judgment or ability to function** **II-3.3= Minor offsite human impacts (e.g. reversible irritation or discomfort)** **Environmental Issues:** **E-3.1= Adverse environmental impairment on and/or offsite—clean up/remediation requiring less than one year** **E-3.2= Potential for incidents that could impact public water works operations** **E-3.3= Minor food chain/terrestrial life/aquatic life impacts**

	Community Relations: **C-3.1= Adverse public or regulatory agency response and local media coverage that is judged to be controllable – Consider the effects of:** **- Minor national media coverage, local media coverage of a week or more** **- Inspections as a result of a complaint** **- Strikes, picketing, protests** **C-3.2= Potential for incidents that could require shelter-in-place for surrounding neighborhoods or businesses**
	Financial Impacts: **F-3.1= Onsite property damage or loss of production revenue of $10 Million to $25 Million** **F-3.2= Minor offsite property damage of $1 Million to $5 Million** **F-3.3= Potential for other expenditures $1 Million to $5 Million**

Overviews of the three **probability** categories are:

- Experience Based—based on the personal experiences of those conducting the review. This is the least accurate method.
- Layers of Protection—counting the independent layers of protection designed to prevent the event from reaching an undesired conclusion. This is more accurate than the Experienced Based method but not as accurate as the Quantitative analysis approach.
- Quantitative—an independent quantitative analysis to determine frequency of occurrence based on known failure rates, probabilities of failure on demand, etc. This is the most accurate of the three methods proposed here.

An example of what the supporting detail for each of the categories might be for a **Level 4 probability** in this risk matrix is as follows. This example is extracted from the full range of five levels of probabilities (Level 1 through Level 5) shown in Attachment A.

4	**Experienced Based: EB-4.1= Event is almost certain to occur in this specific type of process somewhere within the industry during the life of the process, but not necessarily at this location, or;** **Layers of Protection: L4.1= Single level of safeguard <u>plus</u> operator interface, or;** **Quantitative: Q4= Frequency between 1 in 100 yrs to 1 in 1,000 yrs – Between 1×10^{-2} to 1×10^{-3}**

See Attachment A for a complete example of how all of the categories of probability and consequences might be detailed. Again, these are just for illustrative purposes. Each user should define an appropriate number of categories and quantifications that makes sense for their use.

Notice that with this degree of definition and specificity, the risk ranking equation seems more complicated, but the decision logic becomes retraceable:

Risk Rank ~ Probability (Pick the one worse case: 1, 2, 3, 4, or 5)
+ Consequence (Pick the one worse case: 1, 2, 3, 4 or 5)

Each level, 1 through 5, has specific criteria for judgment in both the probability and consequence axes of the matrix. The Risk Rank should be made using the "worse case" from the categories. It is supported and reinforced by lesser cases. See Figure 4.

Risk Ranking Example:
By documenting the decision path using the identifiers provided, the process of converting a hazard to a risk and then ranking it compared to other risks becomes repeatable. Consider this example.

For Consequence:

- The review team determines or estimates that the hazard, if it manifests itself as an event, would likely cause one onsite incapacitating injury (II-3.1). Therefore, for the Injury and Illness consideration, this is a Level 3 Consequence.
- Next the team considers the Environmental impact of the event and determines or estimates it would cause a spill into a containment area (E-1.2) with no other effects. That consequence is a Level 1.
- Then the team considers the Community Relations impact of the event and determines or estimates that there would be minor adverse public or regulatory agency reaction and little media coverage (C-2.2). That consequence is a Level 2.
- Finally, the team considers the Financial Impacts of the event and determines or estimates that there would be onsite property damage/loss of production of less than $10,000 (F-1.2). That consequence is a Level 1.

The highest consequence level, Level 3, is used and the lower consequences become supporting data. The Consequence level would be documented as:
Consequence = Level 3 (II-3.1) + Level 1 (E-1.2) + Level 2 (C-2.2) + Level 1 (F-1.2)
The decision logic, because it is detailed and documented, is retraceable by a different team of people at a different time. It is therefore repeatable.

For Probability:

- The review team has enough information to count independent layers of protection but not sufficient resources to fund a complete quantitative risk assessment. They count independent layers of protection and determine that there are at least two reliable independent levels of safeguards and the failure of one would not allow the event to happen (L-3.1). This is a probability Level 3.

The Probability level would be documented as:
Probability = Level 3 (L-3.1)

Combining the consequence level of 3 and the probability level of 3, the Risk Rank is C, which is a higher risk than A and B, but not as high a risk as D or E.

The Risk Rank equation would then be written:
Risk Rank = C where Probability = Level 3 (L-3.1) and

Consequence = Level 3 (II-3.1) + Level 1 (E-1.2) + Level 2 (C-2.2) + Level 1 (F-1.2)

The Risk Rank of C is supported and reinforced by the additional knowledge that for Environmental, Financial, and Community Relations, the Risk Rank is B. See Figure 4 for a graphical representation.

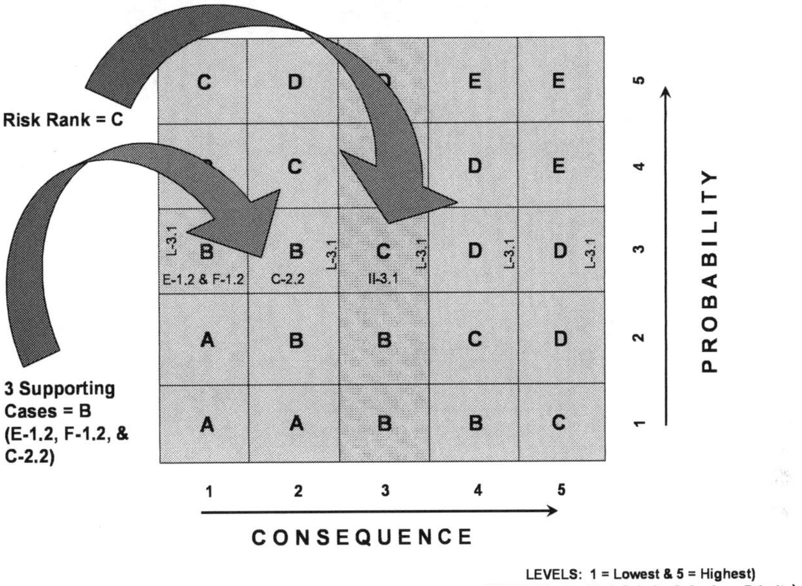

Figure 4

By having the Risk Rank equation document the categories considered and the details used within those categories, the process meets the requirement of **repeatability**. By demonstrating due diligence in defining and using a recognized methodology, delineating the categories and the details therein, the process is **defendable** technically.

Two or More Risks with the Same Ranking:

Ranking risks with this much documented detail has an additional advantage. The process allows prioritization determinations to be made by evaluating the supporting details whenever more than one risk is risk ranked the same. For instance, if two risks are both Risk Ranked C and one is because the Probability and Consequence details are as listed as in the example above,

Risk Rank = C where Probability = Level 3 (L-3.1) and
Consequence = Level 3 (II-3.1) + Level 1 (E-1.2) + Level 2 (C-2.2) + Level 1 (F-1.2)

but the other has the following definition for Consequence,

Consequence = Level 3 (II-3.1) + Level 3 (E-3.2) + Level 2 (C-2.2) + Level 2 (F-2.2) then the second should receive priority over the first because it has a higher risk magnitude in the Environmental and Financial impact supporting data areas. See Figure 5.

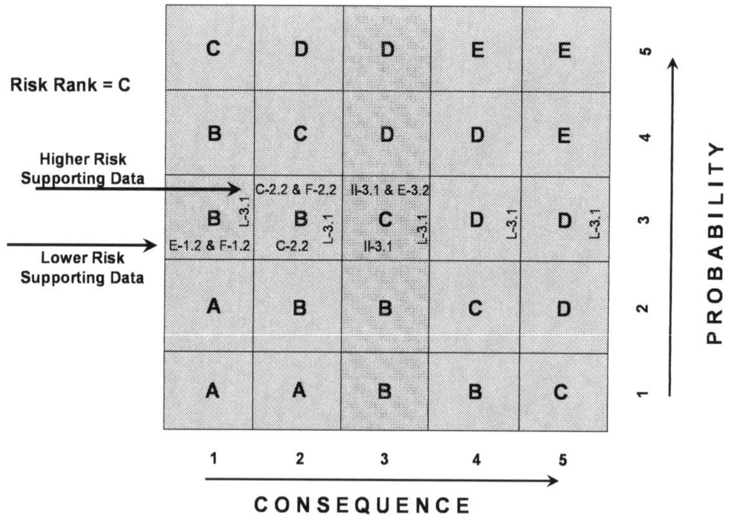

Figure 5

Attachment A is a complete model. It includes an example risk matrix, potential probability and consequence categories and details, a set of risk response categories, and a brief set of use instructions. Each user should carefully consider how to define the categories and details therein for their specific use. Consider that the Risk Ranking Matrix contains **minimum** response expectations. Within these response expectations, the user is expected to develop appropriate action plans and timing for addressing the identified ranked risk. Such actions should satisfy the OSHA's Process Safety Management (PSM) and EPA's Risk Management Program (RMP) requirements as well as the company's standards.

FUTURE POSSIBILITIES

At some point in the not too distant future, the information required for risk ranking may overwhelm a two dimensional matrix and an additional dimension will be required. There are at least two basic ways that a three dimensional matrix may evolve. The first is that it evolves into a cube. This risk cube might have probability and consequence as two of the axes with the third axis being a new variable of concern. For instance, we know that probabilities and consequences are sometimes dependent on or related to exposure time. Therefore, we may want to consider exposure time as the third axis and define it with categories and supporting detail as we did consequence and probability. See Figure 6 for what a risk cube might look like. I leave it to the reader to decide if this is required for their situation and, if it is, to develop the required criteria and supporting details within each level.

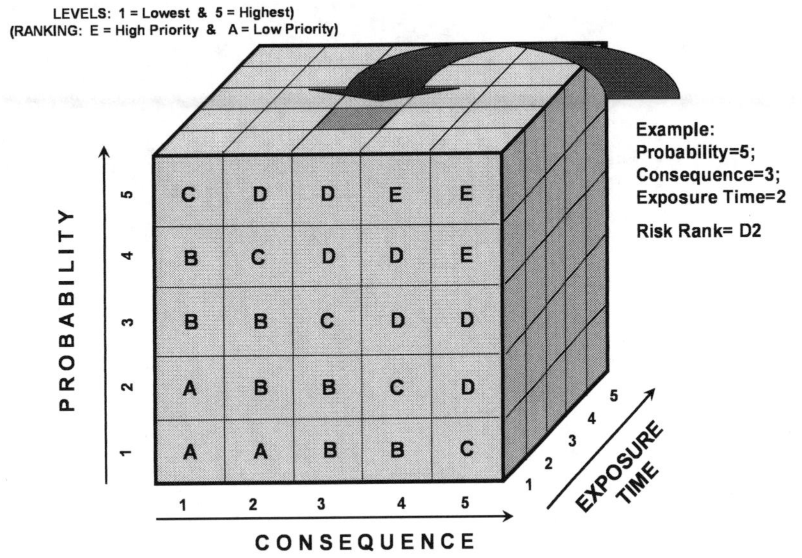

Figure 6

Another route might be to convert each cell within a two dimensional risk matrix to three dimensions by providing a drawer of pull out details to choose from. This approach would allow several different new third axes to exist on the same matrix at the same time. For instance, the Consequence axis of the two dimensional matrix may have exposure time as a third dimension. The Probability side of the two dimensional matrix may have level of accuracy as the third dimension. One or two additional "third dimensions" could be used for each cell within the conventional two dimensional risk matrix. If one is used, each cell becomes a drawer of pull-out details. If two are used, each cell becomes a mini-cube representing a new two dimensional risk matrix. See Figure 7 as a potential example.

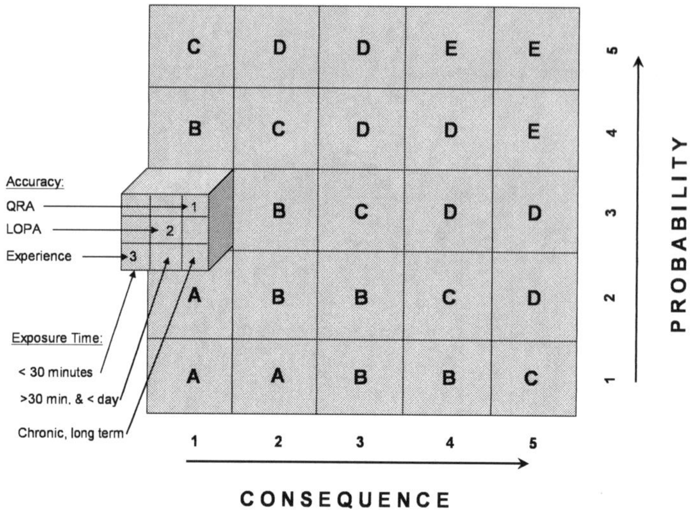

Figure 7

LEVELS: 1 = Lowest & 5 = Highest)
(RANKING: E = High Priority &
A = Low Priority)

Regardless of how the risk matrix evolves, the key criteria is that it should always be **repeatable** (that is, a different group of people with essentially the same information would arrive at essentially the same conclusion) and **defendable** (that is, it has sufficient technology and management systems in place to allow it to be technically correct and defendable).

ATTACHMENT A

RISK RANKING MATRIX USE

SCOPE:

This matrix is a mechanism to characterize and rank risks that are typically identified through one or more multifunctional reviews (e.g., Process Hazard Analyses, PSM Audits, Environmental Audits, Industrial Hygiene assessments, Risk Management Reviews, insurance inspections, incident investigations, facility assessments, etc.)

METHOD OVEVIEW:

1. Identify credible hazards considering the above-described scope.
2. Review each category in all **Consequence Levels** and <u>document all findings</u> associated with that hazard (i.e., identify the potential consequences that would result if the event were to occur). Use the most severe consequence to determine the Consequence Level on the Risk Matrix.
3. Considering the information and resources available, determine the category from the **Probability Levels** that most closely estimates the probability that the identified consequence of concern will occur, (i.e. estimate or calculate the frequency of occurrence). Use that value for the Probability Level on the risk matrix.
4. Apply the consequence number to the X-axis on the matrix and the probability number to the Y-axis to determine the cell into which the risk falls. The letter in the cell is the **Risk Rank** and is to be used to determine the appropriate Risk Response.
5. Refer to the **Risk Response Categories** to determine what level of action should be taken for the identified risk.

Risk Ranking Matrix

					PROBABILITY
C	D	D	E	E	5
B	C	D	D	E	4
B	B	C	D	D	3
A	B	B	C	D	2
A	A	B	B	C	1
1	2	3	4	5	

CONSEQUENCE

LEVELS: 1 = Lowest & 5 = Highest)
(RANKING: E = High Priority & A = Low Priority)

CONSEQUENCE LEVELS (1-5)

5	**Injuries or Illnesses:** II-5.1= More than three onsite permanent incapacitating injuries and/or illnesses II-5.2= Onsite or offsite human permanent incapacitating injury or significant health impact (irreversible reproductive effects, irreversible neurotoxicity, irreversible toxicity to significant body systems, risk of cancer due to known carcinogens) **Environmental Issues:** E-5.1= Extensive environmental impairment on and/or offsite--clean up/remediation not possible; extensive food chain/terrestrial life/aquatic life impacts **Community Relations:** C-5.1= Severe adverse public or regulatory agency reaction and/or major media coverage threatening continued operation of this facility – Consider the effects of: - National media coverage, live TV feeds, three or more days of coverage - Class Action Law Suits, Civil Authority Investigations **Financial Impacts:** F-5.1= Extensive offsite property damage of more than $50 Million F-5.2= Onsite property damage and/or loss of production revenue of more than $100 Million F-5.3= Potential for other expenditures of more than $50 Million
4	**Injuries or Illnesses:** II-4.1= Two or three onsite permanent incapacitating injuries and/or illnesses II-4.2= Potential for onsite or offsite lung dysfunction effects, risk of cancer due to suspected human carcinogens II-4.3= Moderate offsite human impacts (injury or adverse short term health effects) **Environmental Issues:** E-4.1= Serious environmental impairment on and/or offsite--clean up/remediation requiring one year or more; moderate food chain/terrestrial life/aquatic life impacts **Community Relations:** C-4.1= Serious public or regulatory agency reaction -- Consider the effects of: - 60 Minutes type broadcasts - Attacks on products or markets by activists - Potential end-use adverse consequences - Nuisance law suits C-4.2= Potential for incidents that could cause the evacuation of surrounding neighborhoods or businesses **Financial Impacts:** F-4.1= Onsite property damage or loss of production revenue of $25 Million to $100 Million F-4.2= Moderate offsite property damage of $5 Million to $50 Million F-4.3= Potential for other expenditures $5 Million to $50 Million

3	**Injuries or Illnesses:** II-3.1= One onsite permanent incapacitating injury and/or illness II-3.2= Potential for onsite dermal or inhalation sensitization or reversible toxicity that can impair ability to function or judgment II-3.3= Minor offsite human impacts (e.g. reversible irritation or discomfort) **Environmental Issues:** E-3.1= Adverse environmental impairment on and/or offsite—clean up/remediation requiring less than one year E-3.2= Potential for incidents that could impact public water works operations E-3.3= Minor food chain/terrestrial life/aquatic life impacts **Community Relations:** C-3.1= Adverse public or regulatory agency response and local media coverage that is judged to be controllable – Consider the effects of: - Minor national media coverage, local media coverage of a week or more - Inspections as a result of a complaint - Strikes, picketing, protests C-3.2= Potential for incidents that could cause shelter-in-place for surrounding neighborhoods or businesses **Financial Impacts:** F-3.1= Onsite property damage or loss of production revenue of $10 Million to $25 Million F-3.2= Minor offsite property damage of $1 Million to $5 Million F-3.3= Potential for other expenditures $1 Million to $5 Million
2	**Injuries or Illnesses:** II-2.1= Onsite lost time injury or illness or multiple recordable injuries or illnesses II-2.2= Potential for reversible irritation or discomfort II-2.3= No offsite human impacts **Environmental Issues:** E-2.1= Moderate environmental impact on and/or offsite--clean up/remediation is immediate; no lasting impacts **Community Relations:** C-2.1= Minor offsite impact; no lasting consequences (smoke, noise, smell, etc.) C-2.2= Minor adverse public or regulatory agency reaction and little media coverage C-2.3= Inconvenience issues (e.g. , wash and wax cars for dusting) C-2.4= Potential for incidents that require response by local industries and public responders C-2.5= Public notice of excursions **Financial Impacts:** F-2.1= Onsite property damage or loss of production revenue of $1 Million up to $10 Million F-2.2= Minor offsite property damage of $10,000 to $1 Million F-2.3= Potential for other expenditures $10,000 to $1 Million

1	**Injuries or Illnesses:** **II-1.1= At most, one recordable onsite injury and/or illness** **II-1.2= Potential for onsite nuisance effects (e.g. watery eyes)** **II-1.3= No offsite impact**
	Environmental Issues: **E-1.1= Minor to no environmental impact on and/or offsite--no clean up/remediation required** **E-1.2= Spills into containment areas**
	Community Relations: **C-1.1= No adverse public or regulatory agency reaction; no media coverage** **C-1.2= Neighbor inquiries** **C-1.3= Potential for incidents that require only onsite response**
	Financial Impacts: **F-1.1= Onsite property damage or loss of production revenue of less than $1 Million** **F-1.2= Offsite property damage of less than $10,000** **F-1.3= Potential for other expenditures of less than $10,000**

PROBABILITY LEVELS (1-5)

5	**Experience Based:** EB-5.1= Event is likely to occur at this location sometime during the life of this facility EB-5.2= Event has occurred in this specific type of process at another facility using this technology **Layers of Protection**++: L-5.1= Single failure can cause the event L-5.2= Human error(s) alone can cause the event **Quantitative***: Q-5= Frequency of more than 1 in 100 yrs – More than 1×10^{-2}
4	**Experienced Based:** EB-4.1= Event is almost certain to occur in this specific type of process somewhere within the industry during the life of the process, but not necessarily at this location **Layers of Protection**++: L-4.1= Single level of safeguard <u>plus</u> operator interface **Quantitative***: Q-4= Frequency between 1 in 100 yrs to 1 in 1,000 yrs – Between 1×10^{-2} to 1×10^{-3}
3	**Experienced Based:** EB-3.1= Event is likely to occur somewhere within the industry during the life of this general type process **Layers of Protection**++: L-3.1= At least two reliable independent levels of safeguards exist, failure of one NOT allowing the event **Quantitative***: Q-3= Frequency between 1 in 1,000 yrs to 1 in 10,000 yrs – Between 1×10^{-3} to 1×10^{-4}
2	**Experienced Based:** EB-2.1= Similar events are unlikely to occur, but have occurred infrequently somewhere in the world in a similar process **Layers of Protection**++: L-2.1= At least three levels of reliable independent systems are in place, failure of two NOT allowing the event **Quantitative***: Q-2= Frequency between 1 in 10,000 yrs to 1 in 1,000,000 yrs – Between 1×10^{-4} to 1×10^{-6}
1	**Experienced Based:** EB-1.1= Event should not occur during the life of the process EB-1.2 No historical industry experience to suggest that it will occur **Layers of Protection**++: L-1.1= At least four levels of reliable independent systems are in place, failure of three NOT allowing the event **Quantitative***: Q-1= Frequency of less than 1 in 1,000,000 yrs – Less than 1×10^{-6}

++ **Note:** Examples of Protection Layers: preventative measures (check valves, control systems, safety instrumented systems, alarms); administrative measures (training, procedures, PM schedules); mitigation measures (relief valves, secondary containment, fire suppression systems)

* **Note:** Frequency data should only be used when performing a quantitative analysis.

RISK RESPONSE CATEGORIES

Risk Rank

E.

Action:

Immediate mitigation measures are required. If the risk cannot be reduced to a lower category within one week, documented concurrence of executive management is required to continue operations. Perform quantitative risk analysis to verify risk level and assess mitigation opportunities. Opportunities to reduce risks should be reviewed quarterly.

Reporting:

The Vice-President of Manufacturing is to be immediately notified of all items with a Risk Ranking of E and is to approve mitigation steps and timing of action plans. The Risk Management Director is to be notified immediately also. Quarterly status reports to the Manufacturing VP and to the Risk Management Director are required until the issue has been resolved or lowered to at least a Risk Ranking of C. The Manufacturing VP will annually review and reassess risk, mitigation measures and implementation timing for each item with a Risk Ranking of E.

D.

Action:

Verify risk level and document risk studies, potential hazards and relevant risk reducing measures. Facility management is responsible for implementing appropriate measures to reduce risk to at least a Risk Ranking of C within one month. Facility management shall formally inform senior management of the plan and any impediments to meeting the schedule.

Reporting:

The Vice-President of Manufacturing and the Risk Management Director are to be notified within one week of all issues with a Risk Ranking of D. Monthly status reports to the Manufacturing VP and the Risk Management Director are required until the issue has been resolved or lowered to at least a Risk Ranking of C. The Manufacturing VP will quarterly review and reassess risk, mitigation measures, and implementation timing for each item with a Risk Ranking of D.

C.

Action:

Identified risks should be addressed as desirable, but not immediately required, risk reduction opportunities. Additional process risk studies and risk reducing alternatives should be evaluated and implemented where applicable/feasible (cost vs. benefit justified). Document risk studies, hazards, potential risk reducing measures, and conclusions. The Plant Manager is responsible for approving the risk reduction plan, timing and all reporting.

Reporting:

The Plant Manager should notify senior management of the finding and plans to reduce the risk in the quarterly facility status report. Risk Ranked C action items that are not complete and are more than six months old should be separately reported to the VP of Manufacturing and the Risk Management Director.

B.

Action:

Identified risks should be addressed as desirable, but not required, risk reduction opportunities and continuous improvement opportunities. Consider the feasibility and cost vs. benefit of additional risk studies and the development/implementation of risk reducing measures. Document risk studies, hazards, potential risk reduction measures and conclusions reached.

Reporting:

The Plant Manager shall include these items in the annual facility management plan and budget.

A.

Action:

Identified risks should be addressed as continuous improvement opportunities. No mitigation action is required. However, consider the cost/benefit of additional risk studies and the development of risk reducing measures. Document risk studies and conclusions.

Reporting:

None required

References:

- Dumas, Philippe, *Divide and Conquer—A Fundamental Idea*, http://algo.inria.fr/dumas/DC/index_am.html
- Center for Chemical Process Safety, *Layer of Protection Analysis—Simplified Process Risk Assessment,* New York: American Institute of Chemical Engineers, 2001.
- American Institute of Chemical Engineers, *Guidelines for Chemical Process Quantitative Risk Analysis, Second Edition,* New York, 2000.
- Sepeda, Adrian L., *The Development of Risk Profiles and their Value in Making Business Decisions*, CCPS 17th International Conference and Workshop, October 10, 2002.

Implementing a Risk-based Process Safety Management System Because It Makes (Dollars and) Sense

William M. Bradshaw
ABS Consulting
Knoxville, Tennessee
wmbradshaw@absconsulting.com

James R. Thompson
ABS Consulting
Houston, Texas
jthompson@absconsulting.com

ABSTRACT

Process safety practices and formal safety management systems have been in place in some companies for many years. Process safety management is widely credited for reductions in major accident risk and in improved chemical industry performance. Unfortunately, traditional process safety management programs are typically based on implementing a standard set of requirements that, in some situations, may not be enough to adequately manage risk, while in other cases may cause them to "overwork" process safety.

To promote process safety management excellence and continuous improvement throughout industry, the Center for Chemical Process Safety created ***Risk-based Process Safety*** *as the framework for the next generation of process safety management. This approach encourages companies to "consider the risk" when making decisions on how to apply its resources (while of course, continuing to comply with applicable regulatory requirements). It is possible that a company may not be working on the right things, or that it is not doing the necessary things well. Even if the results are satisfactory, resources may be wasted on some tasks that could be done more efficiently or tasks that should not be done at all. This paper examines how to use risk to guide decisions that ultimately steer invariably limited resources in the right direction, helping companies manage both risk and their bottom line.*

1. INTRODUCTION

This paper is largely based on the Center for Chemical Process Safety's (CCPS's) newly published book titled *Guidelines for Risk Based Process Safety*, or *RBPS Guidelines* (Ref. 1). The *RBPS Guidelines* advocate that process safety management systems should be built on four accident prevention pillars:

- Commit to process safety
- Understand hazards and risk
- Manage risk
- Learn from experience

Authentic ***commitment to process safety*** is the cornerstone of process safety excellence. Management commitment has no substitute. Organizations generally do not improve without strong leadership and solid commitment. The entire organization must make the same commitment. A workforce that is convinced that the organization fully supports safety as a core value will tend to do the right things, in the right ways, at the right times, even when no one is looking.

Organizations that ***understand hazards and risk*** are better able to allocate limited resources in the most effective manner. Industry experience has demonstrated that businesses using hazard and risk information to plan, develop, and deploy stable, lower-risk operations are much more likely to enjoy long-term success.

Managing risk focuses on three issues: (1) prudently operating and maintaining processes that pose the risk, (2) managing changes to those processes to ensure that the risk remains tolerable, and (3) preparing for, responding to, and managing incidents that do occur. Most CCPS member companies spend significant time and money in an effort to manage risk. Developing process safety management systems that are tailored to risk helps support incident-free, profitable operations.

Learning from experience involves monitoring, and acting on, internal and external sources of information. Despite a company's best efforts, operations do not always proceed as planned, so organizations must be ready to turn their mistakes – and those of others – into opportunities to improve process safety efforts.

Some companies have implemented very rigorous and detailed process safety management systems, yet they have suffered one or more major incidents. This may be because: (1) employees at the facility do not have an adequate understanding of hazards and risk or how what they do on a daily basis affects risk, (2) a weak process safety culture has allowed management systems to fail, (3) inadequate resources are available to conduct the work activities necessary to protect against a process safety incident (and the culture did not demand that management provide additional resources), or (4) the facility was simply unlucky — even the best process safety management system will not reduce risk to zero. This paper primarily explores the first issue — how improving the understanding of the relationship between tasks that are done each day and their impact on risk improves performance.

2. THE RBPS PREMISE

The RBPS approach recognizes that all hazards and risks are not equal and that the level of resources applied to process safety should be based on hazards and risks. The main emphasis of the RBPS approach is to put just enough energy into each activity to meet the anticipated needs for that activity. In this way, limited resources can be optimally apportioned to improve both facility safety performance and overall business performance.

There seems to be universal support for this approach, particularly with the current demand for lean operations. The corollary RBPS premise is that through better understanding of risk, companies will develop and implement risk-appropriate process safety management systems and drive this understanding throughout the organization. In process safety management, as in almost any other endeavor, improved understanding, particularly improved understanding of cause and effect relationships, leads to markedly improved performance.

A determinant of the rigor needed to effectively manage process safety, which is second only to risk, is process safety culture. The *RBPS Guidelines* define process safety culture as:

> *The combination of group values and behaviors that determines the manner in which process safety is managed. A sound process safety culture refers to attitudes and behaviors that support the goal of safer process operations.*

Culture determines what choices are made by each member of the organization on a daily, if not minute-by minute, basis. In addition, an organization's culture is also a strong determinant of how well the organization nourishes, grows, and enforces its process safety management systems. In providing maximum flexibility in determining the proper mix of work activities to best manage risk, the RBPS approach sets a trap for any organization that has a weak or evolving culture. In the absence of a strong process safety culture, streamlined management systems, along with the notion that everything can and should be optimized, may result in elimination of important but not widely understood process safety management work activities.

The RBPS premise, and the thesis of this paper, is that organizations that understand hazards and risk are better able to allocate limited resources in the most effective manner. Moreover, if individuals within the organization have an understanding of how the actions they perform (or choose to omit) on a daily basis affect risk, and if the organization has a sense of the risk significance of these actions, better risk management decisions will emerge at all levels of the organization. Effectively using risk information to plan, develop, and deploy appropriate management systems that sustain activities critical to properly managing risk helps position the organization for long-term success.

3. BETTER DECISION MAKING AND IMPROVED PERFORMANCE

Thus far we have simply asserted that a better understanding of risk, and in particular a better understanding of risk significance, leads to better decision making. In the area of process safety, this is difficult to prove because catastrophic process safety incidents are rare events. Regardless of risk understanding, an organization may be lucky enough to avoid having a catastrophic process safety incident simply due to chance. Moreover, organizations that do a commendable job of managing process safety may still have a catastrophic incident. No amount of risk understanding, and in fact no process safety management system, can drive the risk to zero.

However, we can draw from other fields, and certainly can point to cases that demonstrate the counter assertion – that lack of understanding has directly led to endorsement of unsafe acts or tolerance of unsafe conditions that ultimately resulted in disaster.

One widely followed endeavor that normally provides immediate and conclusive results is sports. Serious competitors prepare for events both through conditioning and skills training (analogous to the design and training aspects we encounter in industry), trying to determine how their actions may affect the outcome of a contest, and using that knowledge to develop an effective plan for the contest. In addition, most team sports require that plays be designed (a form of procedures), and most serious competitors have a training staff to help prevent or repair injuries (a form of maintenance).

Among two equally prepared teams, the winner is often determined by the quality of decisions and effectiveness of adjustments made during the course of the contest. These decisions are based on an understanding of what will best counter the opponent's strategy, adjustments to playing conditions, and risk management strategies given the current score and the time remaining in the contest.

Another example of risk understanding and the relationship between personal actions and risk reduction, is the use of seat belts while driving or riding in an automobile. Some people wear seat belts because they firmly believe that they save lives. Others wear seat belts because it is required by law. If the laws were to change, or there is a perceived lapse in enforcement, the former group will continue to wear seat belts while seat belt usage among the latter group is likely to decline.

Clearly, process safety is not a game, nor is it as simple as buckling a seat belt. However, we can learn from the effort that coaches invest to instill a high level of understanding in their entire organization or the long-term effectiveness of advertising campaigns that promote seat belt usage. Over the long term, smart choices must be based on the understanding of how what a person does, or chooses not to do, affects the likelihood of a specific desired, or undesired, outcome.

We can also find counter examples in the process safety domain. Each of the 20 element chapters in the *RBPS Guidelines* starts with a summary of a major incident that is relevant to the topic of that chapter. They are designed to grab the reader's interest. Most involve past incidents in the process industries, but there are a few from the commercial aviation industry and space flight as well. Although the summaries are generally too brief to explore fundamental root causes in any detail, a pattern clearly emerges that someone, at some level of the organization, did not understand why something should be done in a particular manner or failed to recognize clear signs that something was amiss, with disastrous results. Examples include:

- A 1974 vapor cloud explosion in Flixborough, England, that killed 28 workers at the facility. The infamous temporary jumper pipe that spanned the gap between reactors 4 and 6 was designed and fabricated by people who did not know how to design large pipes equipped with bellows – the design work for the change was treated more like pipefitting than a large diameter, high pressure piping system design (Ref. 2).

- A 1995 runaway reaction involving a mixture of sodium hydrosulfite, aluminum powder, potassium carbonate, and benzaldehyde exploded and triggered a major fire at a specialty chemical plant in Lodi, New Jersey, killing five workers at the facility. Investigators reported that the immediate cause of the explosion was the inadvertent introduction of water and heat into water-reactive materials during the mixing operation (Ref 3). The same investigation team also determined that (1) an inadequate process hazard analysis was conducted and appropriate preventive actions were not taken and (2) standard operating procedures and training were inadequate.
- A dust explosion at a Corbin, Kentucky, facility that produced fiberglass insulation for the automotive industry. The 2003 incident killed seven workers. Investigators found that the explosion was fueled by resin dust that had accumulated in a production area, which was likely ignited by flames from a malfunctioning oven. Investigators also determined that the company was largely unaware of the hazards of dust explosions and recommended prevention practices and technology available for preventing dust explosions, even though these technologies are listed in numerous industry standards and guidelines (Ref. 4).

Examples are by no means limited to the chemical process industries. Incidents from outside the process industries include:

- A 1979 DC-10 crash at O'Hare airport. The crash killed 271 on board the aircraft and 2 more on the ground. A pylon that held the engine to the wing failed due to a shortcut employed while performing a mandated replacement to some internal bearings. A major U.S. airline instructed its mechanics to remove the engine, together with the pylon, all at one time, using a forklift to hold the engine in place. A failure of the forklift's hydraulic system left the engine unsupported during a shift change, resulting in undetected damage to the pylon. Another major U.S. airline used the same method to remove the engines on planes in its DC-10 fleet, and a total of six DC-10 aircraft were flying with damaged pylons at the time of the crash. Fortunately, the other five were discovered when the Federal Aviation Administration grounded the entire U.S. DC-10 fleet pending inspection of the pylons (Ref. 5).
- A field construction modification to a suspended walkway at a Kansas City, Missouri, hotel introduced a shear/bending stress that was not part of the original design (Ref. 6). Although the change was reviewed by an engineer involved in the construction, he failed to fully investigate the impact of the change in loading. The modified structure could not support the dynamic load that occurred during a special event that was held at the hotel in 1981, causing the walkway to collapse, which resulted in 45 fatalities.

In each case, there was a lack of appreciation for hazards or understanding of how actions workers take (or choose not to take) affect risk. In retrospect, all of these incidents could have been averted if decision makers had the required knowledge of what might go wrong, and they had been supported by a culture that demanded that they stop long enough to figure things out if they lacked sufficient understanding.

Another important RBPS accident prevention pillar is learning from experience, including the experience of others. Once discovered, it is equally important that this knowledge outlive the tenure of the individuals who first discovered it. Traditionally, this was accomplished through life-long employment at a single company and well-engrained, if somewhat informal, mentoring programs. At many facilities this model is failing because individuals tend to change jobs much more frequently than they did 20 years ago. In addition, staffing at virtually all facilities in the process industries has become much leaner. The senior engineer no longer has the time to mentor the new graduate and sees limited incentive for mentoring, given the fraction of new engineers who move on to another position within 2 or 3 years. As competency declines, knowledge-based practices that were once based on solid understanding of the underlying phenomena become rules-based procedures. These rules-based procedures are particularly vulnerable to streamlining, particularly when budgets are trimmed.

4. TAILORING PEFORMANCE TO RISK

In the process industry, relatively few decisions can lead directly to serious accidents such as a crash of a large commercial aircraft, which is likely to result in 300 or more fatalities. Most companies in the process industries could not afford to apply the level of rigor to its safety management systems that are applied in the nuclear and commercial aviation industries and still remain cost competitive. Building a flexible system that is based on risk should help maximize the value obtained from the process safety program. To do this, companies need to ensure that competent persons make risk-informed decisions in a manner that minimizes the chances that they will make a catastrophic error, such as installing an improperly designed jumper line at the Flixborough facility or authorizing the more expedient procedure to remove a DC-10 engine in a single step.

The RBPS subcommittee never intended, nor was it able to provide, a cookbook approach for quantifying risk and mapping it to a set of activities that constitute the optimal system for managing process safety. There is no magic formula or computer program where users input certain data and an optimal management system comes out. Instead, the committee expanded CCPS's original 1989 work from a 12-element process safety management system to the 20-element system (see Attachment) contained in the *RBPS Guidelines*. The committee also incorporated several new elements, such as elements to (1) help maintain organizational competency, (2) ensure reliable operations, and (3) more quickly detect management system lapses using tools that should be more sensitive and responsive than the auditing programs that currently exist at most companies.

The *RPBS Guidelines* contain a description of the essential features of a management system for each of the 20 elements and list a series of work activities for each essential feature, presenting two to five options for executing the work activities that comprise each RBPS element. There is no one right choice for each implementation option, and the RBPS subcommittee does not claim that the list is complete. However, it does provide readers a new perspective. Effectively managing risk, not simply complying with a set of standards or regulatory requirements, should be the goal of any process safety management system. However, process safety management systems (1) cannot be properly designed without a sufficient understanding of hazards and risks, (2) will likely degrade in the absence of a sound process safety culture, and (3) need to take full advantage of learning opportunities to continuously improve.

5. PRACTICAL APPLICATIONS OF RISK-BASED PROCESS SAFETY

The *RBPS Guidelines* are designed to aid process safety professionals in:

- Designing and implementing a process safety management system or element
- Fixing a deficient existing system
- Charting a path for continuous improvement of a mature system.

Readers should also find the *RBPS Guidelines* useful when:

- Searching for ideas to improve the effectiveness of their existing process safety management systems
- Identifying metrics that might be used to provide early warning of management system lapses
- Implementing periodic management reviews to help drive continuous improvement
- Conducting self-assessments of the process safety management programs by comparing the rigor of process safety management work activities at a facility to the range of options presented in the *RBPS Guidelines*
- Providing key personnel with a detailed overview of one or more RBPS elements
- Upgrading process safety overview training and other communication tools

Even if you choose to continue with a compliance-based approach to managing process safety, consider how knowledge of hazards and risks can affect performance. Improving everyone's understanding of the relationship between what they do or choose not to do, and the impact that their actions have on risk, should lessen the frequency and severity of catastrophic process safety incidents.

6. REFERENCES

1. Center for Chemical Process Safety, *Guidelines for Risk Based Process Safety*, American Institute of Chemical Engineers, New York, 2007.
2. Lees, Frank P., *Loss Prevention in the Process Industries, Hazard Identification, Assessment, and Control*, 2nd edition, Butterworth-Heinemann, Oxford, England, 1996.
3. Joint Chemical Accident Investigation Team Investigation Report No. EPA 550-R-97-002, U.S. Environmental Protection Agency, October 1997.
4. *Combustible Dust Fire and Explosions at CTA Acoustics, Inc. Corbin, Kentucky, February 20, 2003*, U.S. Chemical Safety and Hazard Investigation Board, February 15, 2005.

5. NTSB Report Number: AAR-79-17 (NTIS Report Number: NTISUB/E/104-017), American Airlines, Inc., DC-10, N110AA, Chicago International Airport, Chicago, IL, May 25, 1979, adopted on 12/21/1979.
6. Adams, James L., *Flying Buttresses, Entropy, and O-Rings: The World of an Engineer*, Harvard University Press, Cambridge, MA, 1992.

ATTACHMENT

TABLE A.1. Risk Based Process Safety Elements

Commit to Process Safety
• *Process Safety Culture*
• *Compliance with Standards*
• *Process Safety Competency*
• *Workforce Involvement*
• *Stakeholder Outreach*
Understand Hazards and Risk
• *Process Knowledge Management*
• *Hazard Identification and Risk Analysis*
Manage Risk
• *Operating Procedures*
• *Safe Work Practices*
• *Asset Integrity and Reliability*
• *Contractor Management*
• *Training and Performance Assurance*
• *Management of Change*
• *Operational Readiness*
• *Conduct of Operations*
• *Emergency Management*
Learn from Experience
• *Incident Investigation*
• *Measurement and Metrics*
• *Auditing*
• *Management Review and Continuous Improvement*

HOW TO INCORPORATE THE HEALTH, SAFETY AND ENVIRONMENT MANAGEMENT SYSTEM IN QUANTITATIVE RISK ASSESSMENTS

Anna Letícia Barbosa de Sousa[1], Antonio Carlos de Oliveira Ribeiro[2], Fernando José Brasil de Souza[3]

[1] Comissão Nacional de Energia Nuclear- CNEN
Rio de Janeiro- RJ- Brazil
alsousa@cnen.gov.br

[2] BAYER Material Science – HSEQ
Belford Roxo – RJ - Brazil
antonio.ribeiro.ar1@bayer.com.br

[3] Fernando Brasil Engenharia de Segurança Ltda.- RJ- Brazil
Rio de Janeiro- RJ- Brazil
fernandobrasil@uol.com.br

ABSTRACT

Due to the risk involved in some facilities that process hazardous goods, it is mandatory to have a strong management system that includes Health Safety and Environment (HSE) aspects implemented and certified. This is required by shareholders and also by international legislations such as the Directive Seveso II in Europe. Strong evidence in the accident investigation in hazardous process industries shows that 67% of the notified accidents were caused by management failures in HSE. The HSE management has influence in consequence mitigation actions in Quantitative Risk Assessment (QRA), but the risk calculation ist not taken into account. This paper presents a methodology based on audit programs from OHSAS 18001. A model was established to link the correlation between the weight of each of the 16 audited elements and the implementation degree of each element checked in the audit program. The result is transformed into a modification factor to be applied to the failure rate used in the QRA. The impact of this updating in QRA for industrial process plants is showed in a case study.

1. INTRODUCTION

Most of the accident analyses in process plants of dangerous products have showed that the basic causes of these accidents are related to management and organizational factors. These causes meet a detached place when compared to those related to equipment failure. According to Rasmussen 67% of the accidents notified to *European Commission's Major Accident Reporting System* – MARS until 1993 could had their causes distinguished as management failure. This induces us more and more to recognize as true the premise that the Safety Management System (SMS) has an important role to play if a high level in safety is the priority. A great many important chemical and petroleum industries have been adopting health, safety and environmental integrated management systems, which in their turn are also integrated with their administration system.

One way of relating the SMS to a required safety level is to make the performance measures inherent to the administration system to be based on requirements described in a well known guideline, principle or standard. In the development of this work, for instance, we have adopted the directives of the Occupational Health and Safety Assessment Series – OHSAS 18001 which may be considered an efficient tool in auditing activities as well as in symbolizing a powerful indicator of the company's grade of adherence to a model of excellence in SMS.

Fixing the balance of audits through the punctuation of elements taken from an OHSAS 18001 certification process, can generate a correction factor that can be applied in the failure rate used in QRA studies. This procedure can also show itself an efficient tool in the SMS development process and the allocation of resources. It can still demonstrate to the interested parts (workers and their representatives, communities, environmental agencies, shareholders, etc.) the performance's level of influence over the management system in performing risk calculation.

A good management system under the supervision of a systematic audit, which in its turn is based on the OHSAS 18001 certification, will certainly allow us to affirm that the installation has a low risk level. Its implementation will stimulate the development of correction factors correlating the SMS to the QRA.

A likely conception has been used for decades by the insurance market. By the way, the insurance companies have their own methods of punctuation in order to obtain suitable rates for different clients. An other interesting methodology that uses a similar conception is described on API-581 standard (Risk-Based Inspection -RBI).

2. MEASURING SMS PERFORMANCE

When conducting a QRA study using only failure rates taken from generic data bank, there might be classified as equal the risk levels of two installations having completely different SMS performance. In other words, in terms of risk level there might have no difference between an installation highly compromised with SMS, and another one having no obligations or compromises to SMS.

In order that QRA may both provide quantity risk indexes capable of characterizing the plant safety level and be used for the purpose of taking decision on the risk management, it is reasonable to establish an explicit relation between the SMS and QRA.

The audit of SMS like any other management system, normally verifies the existence and implementation of objectives, standards and procedures. The results taken from audits performed in this way are used to validate the efficiency, efficacy and reliability of SMS.

It is advisable to incorporate the influence of the SMS failure rates prior to the mitigation of the consequences, because the SMS influence in probabilistic parameter is clear and well established.

3. THE MODIFICATION FACTOR

The Modification Factor is a way of linking the SMS performance, measured through management audits, with the QRA model.

As a first step, the approach specifies a coefficient (note 1) which represents the influence of each of the 16 audited OHSAS 18001 elements according the QRA parameter. This evaluation is based on the results of an expert's panel with 30 Process Safety specialists, who are members of the Brazilian Association of the Chemical Industries (ABIQUIM). The specialists atribute weights from 1 (minor influences) up to 4 (greater influences). The expert panel results are shown in figure 1.

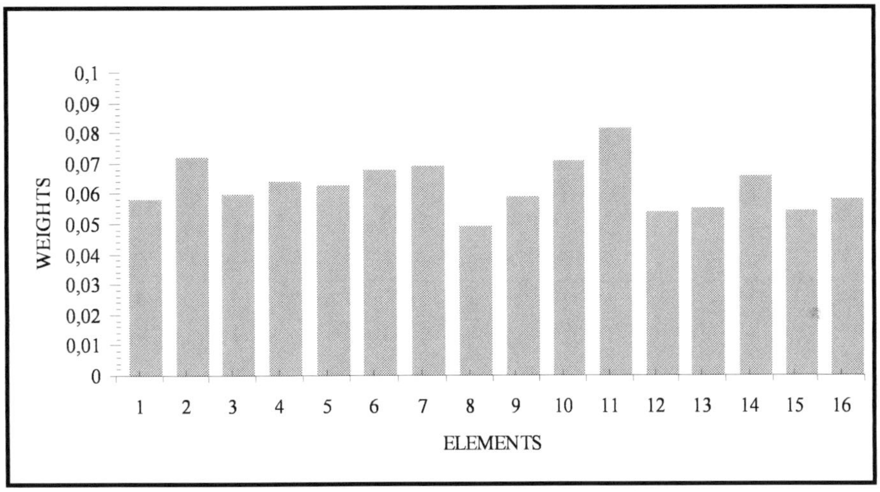

Figure 1- OHSAS 18001 elements weigths

As a second step, a checklist is established for each element through a PDCA (Plan, Do, Check and Act) evaluation process; then the OHSAS 18001 elements are assessed so as to give the implementation grade of each of the 16 elements.

The punctuation goes from 0 (not attempted) to 10 (fully attempted). As an example, the PDCA of the Element 2 – Risk/Hazards and Environmental Aspects/Impacts Identification and Element 7 – Training, Awareness and Competence are given in Table 1 and Table 2, according to SGS, ICS Portugal, Interpretation of OSHAS 18001 [7].

Table 1 – OHSAS 18001 elements implementation grade (best case scenario evaluation)

Element 2 – Risk/Hazards and environmental aspects/impacts identification	Evaluation	Comments
Is there a procedure to identify hazards for workers' health and safety related to the company activities?	10	*Procedure in place*
Does this procedure also involve necessary controls and measures related to the identified hazards?	10	*Yes, fully*
Occasional and normal activities are also considered?	7	*Some normal activities not considered*
Does the methodology include installations, process and contractors?	8	*Not all contractors are included*
Does the methodology identify the risks that should be reduced or controlled?	10	*Yes, fully*
Are control measures established for unacceptable risks?	–	*Not applied*
Does this methodology give sufficient information to develop new and existing process controls?	8	*Some areas are not fully attempted.*
Do company goals take into account the results of identified hazards and risks and control measures?	0	*No evidence*
Is there a follow up for the required action implementation?	8	Yes, but not enough assured.
Total	61 (76,25%)	

Table 2 – OHSA 18001 element implementation grade (worst case scenario) evaluation

Element 7 – Training, awareness and competence	Evaluation	Comments
Is there a methodology to identify the awareness and competence in OH&S (Occupational Heath & Safety) throughout the different hierarchy levels in the company?	0	*No methodology available*
Is there a methodology to assure that all workers have specific formation in theirs activities, risks associated, correlated safety precautions and procedures to follow?	5	*There are some risks not identified in the procedures.*
Does the existing procedure include recycle training for any new job activity in the company in a systematic way?	8	*Not all activities are included.*
Does the training program assure the workers the capacity to identify hazards and risks as well as the tools to keep them under control?	8	*Contractors are not included.*
Does this program assure that contractors' temporary workers and visitors are informed about the risks associated with theirs activities?	0	*No*
Does the training program have a methodology to measure its efficiency?	5	*Some trainings don't have evidence of evaluation*
Is there a methodology to solve any worker formation deficiency?	0	*No*
Are there records of individual training?	10	*Yes.*
Total	36 (45%)	

This evaluation must be developed for each OHSAS 18001 elements in order to achieve the implementation grade value, necessary to have note 2 as seen in Table 3.

Table 3 – Definition of each OHSAS 18001 elements implementation grade

Note 2	Classification	Punctuation definition
-1	very good	The implementation grade is greater than 75%.
0	good	The implementation grade is between 50 and 75%.
1	poor	The implementation grade is lower than 50%.

Combining note 1(expert judgments) with note 2 (implementation grade) in table 4 results in a modification factor for the SMS influence in the failure rate, defined by equation 1 and latter used in equation 2.

$$m = \sum_{i=1}^{16} note1 * note2 \qquad \text{Eq-1}$$

Table 4 – note 1 and note 2 combination matrix for all OHSAS elements

	Element	Note 1	Note 2
1	-Occupational health and safety policy.	0,06	-1
2	-Identification of hazards, assessment risks, and the implementation of necessary control measures.	0,07	-1
3	-Identification and compliance with legal and other OH&S requirements.	0,06	0
4	-Documented occupational health and safety objectives at each relevant function and level within the organization.	0,06	1
5	-Management program to achieve the objectives.	0,06	0
6	-The designated responsibility and authority of personnel to perform and verify activities having an effect on the OH&S identified risks.	0,07	-1
7	-Training, awareness and competence program.	0,07	1
8	-Consultation and communication.	0,05	-1
9	-Documentation.	0,06	0
10	-Documentation and data control (Management of change).	0,07	-1
11	-Operational control (Equipment inspection, and corrective, preventive and predictive maintenance program).	0,08	-1
12	-Emergency preparedness and response.	0,05	0
13	-Performance measurement and monitoring.	0,06	0
14	-Accidents, incidents, non-conformances and corrective and preventive actions.	0,07	-1
15	-Records and record management	0,05	1
16	-Audit	0,06	-1

So as to adjust the failure rate to SMS performance, a model based on Papazoglou (1999) and in the API – 581 standards is proposed.

$$\log f_{\text{mod}} = \log f_0 + m \quad \text{Eq-2}$$

f_{mod}	Modified failure rate
f_0	Failure rate from generic data bank
m	Modification Factor from equation 1

So as to harmonize the proposal model it is necessary to keep in mind and answer the following considerations:

- ✓ This factor derives from the results yielded by evaluating a facility or an operating unit SMS. Different practices within units at a facility might create differences in the SMS evaluation factors between the units. However, within any of these studies, the management of the system evaluation factor should be the same. The factor is applied equally to all equipment items [1].
- ✓ An average plant that score 50% on the SMS evaluation will have a 0 as note 2
- ✓ When all OHSAS element implementation grades are greater than 75% (excellence in SMS), it results in a reduction of the failure rate ten times greater than usual.

4. CASE STUDY

The application of the procedure to an LPG storage facility will allow the validation of the method.

4.1 Description of the Installation

A liquefied petroleum gas (LPG) storage facility with 18 tones of capacity. The gas is used as fuel for ovens and welds. The LPG is transferred to a cylinder tank from truck storage, unloaded to a vaporizer and from the vaporizer to its final destination.

The operation of the storage occurs in three stages:

1 – The LPG is unloaded from truck through a flexible hose using the truck's pumps. Each shipment operation has a 2 hour duration and it happens each 30 days.

2 – The LPG is stored at 6 kgf/cm^2 and 25°C. The tank is equipped with two safety valves, which open in case of over pressure.

3 – The liquid LPG (6 kgf/cm^2 and 25°C) is fed in the vaporizers. The gaseous GLP (6 Kgf/cm^2 and 30°C) flows to the tank lung through a valve that reduces the pressure to 2.5 kgf/cm^2. The gaseous LPG feeds the areas of process through line of 6" (2.1 kgf/cm^2 and 25°C).

4.2 QRA Methodology

The frequency of LPG release to the environment, its subsequent events and the corresponding consequences were estimated via the QRA methodology which consists of the following basic steps.

- Hazard identification;
- Accident sequence modeling;
- Consequence and vulnerability assessment;
- Frequency assessment;
- Risk calculation and assessment.

The following accidental hypotheses for a significant leakage of LPG had been raised in the hazard identification stage:

- Disconnection of flexible hose during shipment operation (failure in connection);
- Breakage of flexible hose caused by an unexpected movement of the truck during shipment operation;
- Material failure of flexible hose;
- Catastrophic failure of LPG tank;
- Pipe breakage in the loading section caused by external impact;
- Pipe breakage during the unloading to process areas caused by excess of vibration (pump discharge);
- Pipe break in the unloading to process areas by external impact.

The consequence, vulnerability and frequency assessment were performed for all listed above accidental hypotheses. The individual risk calculation was raised for three cases of SMS performance:

1 – For the best SMS performance by using a modification factor corresponding to implementation level above 75% for all elements of OHSAS 18.001 a procedure that results in a reduction equal to 0,1;

2 – For the measured performance of SMS based on OHSAS 18.001. In this case study the model gives a reduction equal to 0,46;

3 – Without SMS influences – failure rates from generic data bank.

4.3 Results

Table 5 – Accidental hypotheses for individual risk calculation

Accidental hypotheses description	Failure rate from data bank traditional risk analysis (oc./year)	Modified failure rate from measured performance of the SMS (oc./year)	Modified failure rate optimal performance of the SMS (oc./year)
Disconnection of flexible hose during shipment operation (failure in connection);	1,37E-05	6,32E-06	1,37E-06
Breakage of flexible hose by truck unexpected movement during shipment operation	2,7E-06	4,62E-04	2,7E-07
Material failure of flexible hose;	1,37E-05	6,32E-06	1,37E-06
Catastrophic failure of LPG tank;	9,55E-05	4,42E-05	9,55E-06
Pipe breakage in the loading section by external Impact;	3,87E-03	1,79E-03	3,87E-04
Pipe breakage in the unloading to process areas section by vibration excess (pump discharge);	3,87E-03	1,79E-03	3,87E-04
Pipe break in the unloading to process areas section by external impact.	3,87E-03	1,79E-03	3,87E-04

Table 6 – Individual iso-risk curves extension

Risk level (Fatalities/year)	Distance from facility traditional risk analysis (m)	Distance from facility measured performance of the SMS (m)	Distance from facility optimal performance of the SMS (m)
1,0 E-04	41	41	21
1,0 E-05	61	53	50
1,0 E-06	194	64	55
1,0 E-07	266	252	195
1,0 E-08	352	303	274

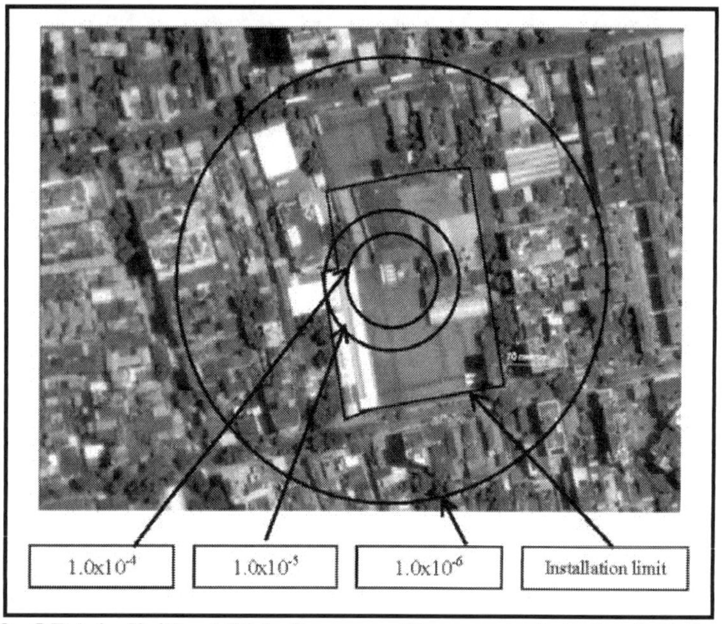

Figure 2- QRA individual Risk Curves with generic data bank failure rates

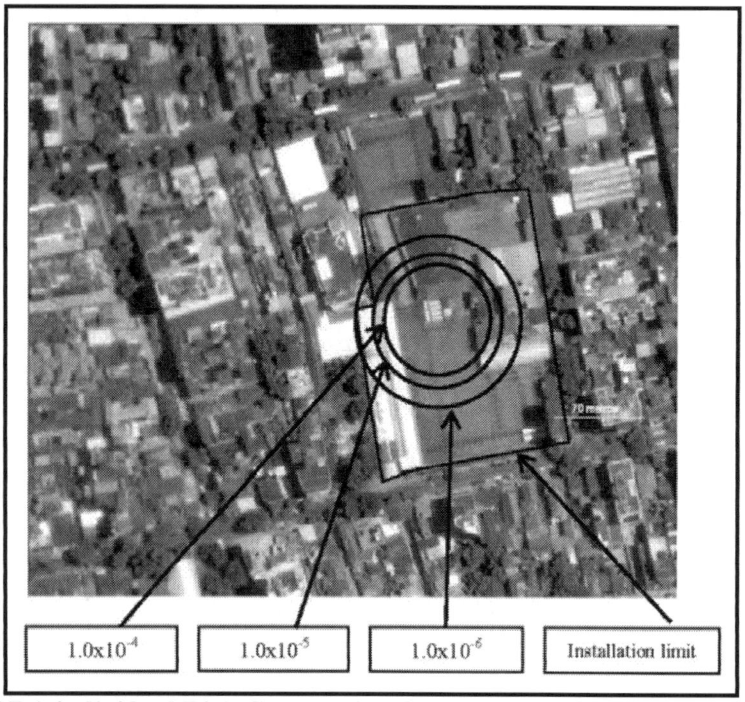

Figure 3- QRA individual Risk Curves with SMS performance measured correction factor applied to failure rates (case study presented in this paper)

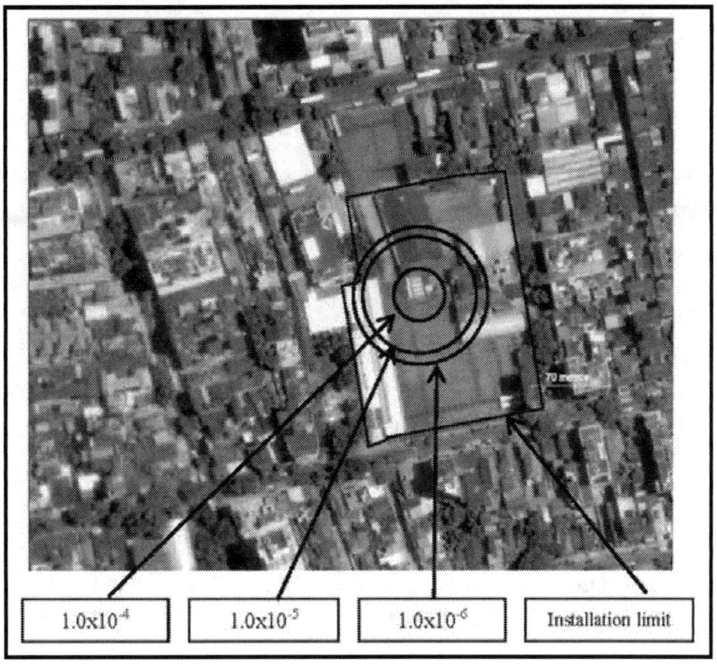

Figure 4- QRA individual Risk Curves with SMS correction factor applied to failure rates (maximal reduction factor, excellence in SMS)

5. CONCLUSION AND RECOMMENDATIONS

The use of a modification factor applied to failure rates in QRA studies is not new. There are some industrial Standards for those factor applications such as MIL Std 217 for electronics components, NUREG 1100 for nuclear power plants and the methodology proposed by Ribeiro (2006) for process plants. All these publications have the same limitation. They need a historical failure data bank from the plant. And effort is required to generate this data bank. On the other hand, the here proposed methodology can be easily applied in any industry that has an auditable SMS.

The incorporation of SMS evaluation in QRA is a very important way to demonstrate the installation safety culture as well as a detailed view of its own risks. The proposed methodology in this work uses a usual and available tool in the market: The OHSAS 18001 certification program. About OHSAS 18001: it is very important that the final audit report has a detailed implementation grade for each element. In this way subjectivity can be reduced in the results, providing a more reliable measurement.

In some countries like Brazil, Netherlands, and England, the environmental licenses for hazardous process plants are based on risk acceptance criteria using QRA assessment. Our intention with this work is to give a contribution to reduce the gap between QRA numeric results and the operator safety cuture.

6. REFERENCES

1. API, (American Petroleum Institute), Risk Based Inspection Base Resource Document, API 581, 2001.
2. DEMICHELA, M., PICCININI, N. How the management aspects can affect the results of the QRA. Journal of Loss Prevention in the Process Industries, 19, 70–77, 2006.
3. DEMICHELA, M., PICCININI, N., & ROMANO, A. Risk analysis as basis for safety management system. Journal of Loss Prevention in the Process Industries, 17/3, 179–185, 2004.
4. MITCHINSON, N., & PAPADAKIS, G. A. Safety management systems under seveso II: Implementation and assessment. Journal of Loss Prevention in the Process Industries, 12(1), 43–51, 1999.
5. PAPAZOGLOU, I. A., & ANEZIRIS, O. On the quantification of the effects of organizational and management factors in chemical installations. Reliability Engineering Systems Safety, 63, 33–45, 1999.
6. PAPAZOGLOU, I. A., ANEZIRIS, O. N., POST, J. G., & ALE, B. J. M.. Technical modelling in integrated risk assessment of chemical installations. Journal of Loss Prevention in the Process Industries, 15, 545–554, 2002.
7. SGS, ICS Portugal, Interpretação da OHSAS 18000.1 sem acidentes, volume 1.1, 2003.
8. SMITH, A J. The development of a model to incorporate management and organizational influences in quantified risk assessment. HSE Contract Research Report No. 38, 1992.
9. RIBEIRO, A., MELO, P.F., Impact of failure data especialization in quantitative risk assessment of process plants. 21st Annual CCPS – AIChE International conference on process Safety Challenges in a Global Economy, 2006.
10. VAN VUUREN, W. Cultural influences on risks and risk management: Six case studies. Safety Science, 34, 31–45, 2000.

Excellence in Stationary Equipment Reliability

By

F. Walter Pinto
Lyondell Chemical Company

Abstract

This paper describes the elements for a successful stationary equipment reliability program in a petrochemical facility. The paper addresses management systems, engineering practices, preventive/predictive maintenance/inspection systems, performance metrics and resources. The paper shares the stationary equipment reliability program at Lyondell Chemical Company / Equistar Chemicals, LP a number of best practices developed as part of the reliability program. Also, some success stories and lessons learned are shared. A model stationary equipment reliability program outlined in Attachment 1.0 can be used to evaluate the reliability program at any site. A list of abbreviations used in this paper is provided in Attachment 8.0.

Introduction

A plant engineer is driving to work in the morning. He sees the emergency flare light up and says to himself, "there it goes again!! We must have had another major leak due to Corrosion-under-Insulation (CUI)." Later, he finds that it was indeed a leak due to CUI, and the associated cost was in excess of two million dollars in lost opportunity and hydrocarbons flared. I am sure all of us have experienced repeated equipment failures and said to ourselves, "Why can we not learn from these incidents and prevent them from occurring again?" Some companies do and some don't. The only way unplanned equipment failures can be avoided is by having a sustained "reliability culture" and not by "bandage approach."

This paper focuses on stationary equipment reliability. However, the principles can be applied to other disciplines. A number of papers have been published in the area of reliability. Science and the tools in this area are well known and available. The challenge always is in **implementation** because it requires resources. A reliability culture can only exist if it has "reliability lobbyists" to make a constant case with the decision makers.

Reliability at Lyondell/Equistar is defined as "the availability of operating units on demand." The focus is on "excellence" not "compliance." The reliability program covers all equipment not just "process safety management (PSM) covered" equipment. We believe that reliability drives excellence in safety, cost, quality and environmental performance. Our experience shows that all roads lead to reliability. The most reliable plants have the lowest maintenance cost and excellent safety performance. Attachment 2.0 illustrates this experience within Lyondell/Equistar. The first graph shows the relationship between reliability and safety performance at a site. The second graph shows the relationship between reliability and maintenance cost performance at another site. Reliability is a competitive advantage.

Why Focus on Stationary Equipment?

Stationary equipment includes equipment types such as pressure vessels, heat exchangers, piping, storage tanks, valves, pressure relieving devices, boilers, furnaces/heaters and structures. In a typical refinery or petrochemical facility a significant portion (generally more than 50%) of the capital/maintenance cost and reliability events or failures are associated with stationary equipment. The risks from stationary equipment are highest compared to other equipment types because of the shear number of equipment and quantity of stored fluid. A significant percentage of industry incidents that resulted in injuries or asset loss also have been associated with stationary equipment.

In cases like heat exchanger tubes or expansion bellows, the critical components are made of thin sections by design. Stationary equipment is often subjected to extreme operating conditions and varying damage mechanisms. In most of the cases, the damage is not immediate. A good example is CUI damage. How many times have you seen insulation cut open during maintenance or a turnaround and left open, exposing the previously covered pipes to weather conditions for months with no protection? Operations and maintenance personnel forget that rain water has made the insulation wet, thus creating a perfect condition for CUI. Of course, in addition to CUI, the insulation is rendered useless due to increased conductivity and heat loss. Why is this allowed? This behavior is accepted because the leak is not immediate. The consequences will only be realized after a few years, but by then everyone has forgotten about how the water got inside the insulation.

Unlike machinery, stationary equipment is not fitted with "bells and whistles" for monitoring conditions on-line. Stationary equipment conditions are generally monitored at specified intervals, depending on damage mechanisms, and even then only a small percentage of area generally is inspected. This makes it especially critical to have a robust reliability program. In best-in-class organizations, stationary equipment reliability always leads overall reliability.

Focus Areas of Stationary Equipment Reliability

There are multiple approaches for designing a reliability program. We believe there are seven focus areas for the stationary equipment reliability program. They are:

1. Management Systems
2. Engineering Practices
3. Reliability Practices
4. Operations Discipline
5. Implementation Discipline
6. People
7. Management Support

One can compare stationary equipment reliability to a knock-out punch. The first five focus areas represent the fingers of the fist. None of these focus areas can be effective without a high-performing trained workforce. The people represent the palm that holds all the fingers. Of course, one can not deliver a knock-out punch without the supporting force delivered by the arm. Similarly, reliability programs can not be successful without a strong management support. Each of these focus areas is discussed in detail in the following sections. The scenarios given are fictitious but based on industry experience.

Management Systems

There is a major fire due to a leak caused by CUI. During the investigation it was found that this line not only was overdue for inspection, but that it had never been inspected. This was not an oversight, but a case of "budget blinders." The inspection would have required scaffolding, and the operations superintendent did not want to spend money on scaffolding. After the fact, the plant manager asked why he had not been told about this past-due inspection.

This type of situation can be avoided by establishing strong management systems. Management systems are procedures that define and control the work process and define roles and responsibilities. We suggest the following six steps for implementing effective management systems:

1. **Reliability Beliefs and Principles:** These define the vision and strategic direction for the reliability program and set the foundation for it. Executive management must endorse this document and continually communicate their belief in these beliefs and principles. Although this document might contain "motherhood and apple pie" language, it is critical to have this document as a guiding principle. Lists of reliability taboos also help engrain the reliability culture.

2. **Performance Standards:** These are high level standards that define the required management systems. An example of a requirement is "A management system for timely notification to site management of past due inspections and recommendations." We suggest these documents be approved by an executive officer, preferably the chief operating officer.

3. **Inspection and Maintenance Standards:** These standards define the technical requirements for inspection, repair and maintenance of existing equipment. These standards should have a single owner with responsibility and accountability for technical content. A management system should be established for processing and approving addenda and deviations.

4. **Inspection and Maintenance Procedures:** These procedures define workflow, roles and responsibilities and are controlled and approved by the plant management. These procedures are consistent with the requirements of Performance Standards. A flow chart format is preferred for specifying workflow. Suggested inspection and maintenance procedures are included in Attachment 3.0.

5. **Key Performance Indicators:** Establishing key performance indicators (KPI) is critical for measuring current performance and indicating opportunities for improvement. These are performance-related actionable metrics designed to improve reliability and cost objectives. Often organizations report KPIs but never analyze and act upon them. There is absolutely no benefit in reporting KPIs if improvement projects are not implemented.

Some of the KPIs used at Lyondell/Equistar in the stationary equipment area are:

- Past due inspections
- Past due recommendations
- Number of pressure boundary leaks
- Relief valve pre-test failures
- Weld reject rate
- Receiving inspection reject rate

Internal and external benchmarking is performed to evaluate staffing and cost structure. The main sources for external bench marking are Solomon Study results and the API Inspection Survey.

Attachment 4.0 shows the reduction in pressure boundary leaks due to CUI at a facility in Europe. Investment for CUI remediation was increased in early years, resulting in dramatic reduction in the leaks. Note that a constant $500M is spent every year for CUI inspections and maintenance.

6. **Compliance Audits**: Audits are a mechanism to determine performance of management systems and programs. These audits can be viewed positively or negatively. At Lyondell/Equistar, these audits are considered opportunity for improvement. Rigorous audits are performed at a fixed interval (generally every three years). Auditors are company employees but are from outside the site being audited. A detailed audit protocol exists. The auditors are trained in the audit protocol and are knowledgeable in the company performance, inspection and maintenance standards. The auditors verify existence of management systems to meet the corporate requirements. Performance is verified through a vertical slice audit. Evidence is required for verification.

   The findings and observations are documented and reviewed with plant management staff at the end of the audit. Inspection management audits last about two weeks. The action items are monitored by management. Closure is required within a fixed time frame (usually one year).

Engineering Practices

The unit experienced numerous outages because of heat exchanger leaks. This unit was constructed just five years ago and built in accordance to the contractor standards. An investigation indicated a series of problems. The design did not consider minimum velocities in the exchanger tubes and a number of process exchangers were designed with welded U-tube bundles, requiring cutting the bundle for removal. Moreover, the contractor standard was used only as a guideline, a number of deviations were allowed.

This type of situation can be avoided with the implementation of strong engineering practices. The two elements of engineering practices are:

1. Engineering Standards
2. Engineering Data Management

1. **Engineering Standards:** Every organization must have Engineering Standards that define design, fabrication, inspection, testing, shipping and field construction requirements. Reliability requirements are built into the Engineering Standards. The requirements of these standards are mandatory. This is a big challenge and can be perceived as introducing bureaucracy and stifling creativity. However, experience indicates that a properly managed program can cut debate, eliminate costly mistakes and implement lessons learned. To make the standards program successful, manufacturing and engineering senior management must mandate that the standards be followed.

   The Engineering Standards have a single owner who is held responsible and accountable for technical requirements. A process exists to capture plant or project- specific technical requirements in the form of addenda to the Engineering Standards. There is a process for managing deviations to the Engineering Standards. The addenda and deviations are approved by the standards owner. The standards are administered by a dedicated group and are easily accessible through the company intranet. Standards are constantly evolving documents. Lessons learned from every application is captured, evaluated and implemented by the standards owner.

   There are significant benefits in having a Global Piping and Valve Specifications as part of Engineering Standards. Cost savings can be realized by streamlining procurement of piping components, eliminating duplicate efforts at multiple sites, engineering efficiency and minimizing errors through automation. Of course, there are challenges for implementing new specifications at an existing site. The best way to accomplish this is to develop a cross-reference list between the old and new piping and valve specifications. Having a cross-reference list will avoid revising P&IDs to reflect new specification numbers. Any new lines added will have the new specification numbers specified in the P&IDs.

2. **Engineering Data Management:** Plants tend to have design-specification data at multiple locations such as Computer-Based Maintenance Management System (CMMS), spreadsheets and Inspection Data Management System. Quite often the data is conflicting. Best-in-class companies establish master systems for each data or document type. Other systems copy data for display from the master database. Data can be altered only in the master database. Manufacturing mission-critical documents such as P&IDs, PFDs, equipment drawings and design calculations are stored in a central computer-based system. This allows employees and contractors efficient access and reuse of information over the equipment life cycle. Equipment purchase specification data are stored in a query-able format in a master database. Engineering data are loaded into the system at the end of each project and turnaround. P&IDs, PFDs, equipment lists, equipment drawings and datasheets are updated promptly when modifications or repairs are made to the equipment.

Reliability Practices

Within two months after a turnaround, three critical heat exchangers experienced tube leaks, resulting in plant shutdowns. These exchanger tubes were tested during the turnaround by a contract crew. The leaking tubes were pulled and split, and pits were measured. All the leaking tubes had multiple pits with wall loss in excess of 75%, including the holes where the tubes leaked. The testing crew had tested these tubes and concluded the tube wall loss to be less than 20%. The technicians had a level 2 certificate for performing tube tests from their employer. Serious questions were raised on the rest of the test results. Later, an exchanger tube testing qualification program developed within the company found that the crew was not competent to perform tube testing.

These types of incidents are common in organizations where good reliability practices lacking. The three key elements of reliability practices are:

1. Inspection Management
2. Focused Reliability Improvement Programs
3. Failure Reporting and Investigation

1. **Inspection Management**: This is a critical element and helps monitor the condition of each piece of equipment. The goal is to detect deterioration prior to actual failure. The following seven steps are suggested to establish an inspection program:

   a. Establish a corrosion manual. This is the foundation for the inspection program. It is developed with the input of process, inspection and metallurgical engineers. Operators and engineers are trained on the content of this corrosion manual. A corrosion manual contains:

      - Process description for each unit and section.
      - Corrosion table listing product characteristics, process conditions, corrosion mechanism and failures.
      - Failure history at similar process plants and industries.
      - Current metallurgy and improvements made to the metallurgy over the plant life.
      - Key corrosion variables and operating envelope. This requires investigation when operating conditions deviate from the operating envelope.

   b. Perform risk analysis to determine inspection priority and develop an inspection plan. Quantitative risk analysis in accordance with API 580/581 is suggested for pressure vessels, heat exchanger shells and piping circuits. API 580, Appendix O, can be used with some enhancement for storage tanks. It is strongly recommended that a plant inspector or engineer perform the Risk Based Inspection (RBI) analysis. Many a times, due to resource constraints, plants hire outside consultants to perform RBI analysis. RBI analysis by outside consultants can be successful only if plant inspection personnel are fully engaged and verify input and output data.

      At Lyondell/Equistar, the RBI results are being used to identify equipment for CUI remediation, determine predictive inspection scope for turnarounds and focus resource spending on damage source and on high-risk equipment. In a typical petrochemical facility more than 80% of the failures are due to external corrosion; yet, in the past the facilities have spent more than 80% of the total inspection cost in performing internal inspection. In non-corrosive services, significant cost savings can be realized by performing external inspection in lieu of internal inspection.

A qualitative or semi-quantitative tool works well for evaluating the risk in heat exchanger bundles. The tool used at Lyondell/Equistar is called the Retube Analysis Tool (RAT). This is a point-based system with 0 and 16 being the lowest and highest risk, respectively. Four risk factors are used in the matrix. They are Tube Age Factor, Remaining Life Factor, Production Criticality Factor and Service Factor. A maximum of 4 points can be obtained for each factor, with the higher numbers indicating worsening conditions, and the lower numbers indicating better conditions.

The inspection and retube priorities are based on the total number of points. The following are the three risk categories:

Risk Category	Total Points
Retube	11 – 16
Priority Inspection	8 – 10
Normal Inspection	0 – 7

Reference (1) provides more information on the heat exchanger reliability program at Lyondell/Equistar.

Similarly, a point-based qualitative tool works well for relief valve inspection and testing. This tool is designed to include past inspection and test data to help determine inspection interval.

c. Ensure that personnel performing NDE and inspections required by API standards are qualified. Contract inspection and NDE personnel must be qualified to the industry-accepted standards. It is recommended that an internal qualification program be developed to qualify technicians for ultrasonic shear wave inspection, heat exchanger tube testing, acoustic emission testing and other high technology inspection methods. A core group of contract personnel should be identified for these inspection methods.

   At Lyondell/Equistar, a program was developed to qualify tube-testing NDE technology, hardware and technicians through performance tests. The purpose of the program was to test, rank and use only the top tier technologies, hardware and technicians to obtain reliable tube test data during turnarounds. A mock exchanger was fabricated using the most common tube sizes and materials found in Lyondell facilities. Each tube had a known set of flaws. The types and locations of flaws simulated actual situations. In addition, tubes removed from plant exchangers with actual flaws were used to satisfy quantitative needs and to simulate the liftoff effect that exists in dirty tubes. Random placebo tubes (tubes with no defects) were installed to further validate operator performance. More than 50% of the ASNT level 2 and 3 operators were disqualified using this qualification process. References (1) and (2) provide more information.

d. Perform inspections on a timely basis. This is where the "rubber meets the road." To be successful in this activity, follow these steps:

- Develop the following year unit inspection plan in the 2nd quarter of the previous year. This will help the maintenance department budget cost of maintenance support for inspection activities. Operations can plan to make specified equipment available.

- Publish 90-day look-ahead inspection activity each month to plant management and other stake holders. This will ensure timely planning of inspection activities.

- Develop the inspection plan and checklist for performing and documenting inspections.

- Publish past-due inspection report at a prescribed regular interval to management to focus attention to inspection back log.

e. Evaluate inspection results (condition assessment). It is the responsibility of the unit inspector to review the inspection results to determine if findings and observations meet the acceptance criteria. Generally, this exercise is limited to API 510, 570, 653 and API 579 level 1 methods. Engineering assistance will be required to perform level 2 or 3 fitness-for-service evaluation. Each company should decide at what damage level the inspection personnel notifies management and if a more formal process is required.

   At Lyondell/Equistar a Mechanical Integrity Risk Assessment (MIRA) program is used to justify continued operation of equipment with damage failing level 2 assessment. Management is notified immediately and resources are allocated. A multi-discipline team is assembled to perform the MIRA. Technical evaluation includes review of failure modes, fitness-for-service evaluation, potential risks, justification for continued operation, action plan to mitigate risk and safety, operational and maintenance issues. The MIRA requires approval by the plant manager. Corporate specialists have expertise in API 579 methods, stress analysis, finite element methods and fracture mechanics for evaluating damage. Several computer programs are available for stress analysis (static and dynamic) of pressure equipment and determining stress intensities at the flaws.

f. Set corrective recommendations and an action plan. Evaluation of inspection information often results in recommendation to remediate deficiency. Management of these recommendations is critical to the success of a reliability program. Integral to this system is the management of inspection data. Inspection activities generate the following documents and data:

- RBI analysis data including risk category, inspection due date and required type of inspection effectiveness.
- Unit and equipment inspection plan for executing inspections.
- Annual and 90-day look ahead plans.
- Inspection plan date and notification number for maintenance support.
- Inspection report documenting findings and observations.
- Thickness data and corrosion rate.
- Condition assessment and fitness-for-service evaluation reports.
- Recommendations, associated maintenance notification number and closure report.

The Inspection Data Management System at Lyondell/Equistar allows management of the above data in one application. The hierarchy used for data management is site, plant, unit, area and equipment. Under each equipments folder are folders for equipment data, RBI data, inspection plans, inspections results and recommendations. The application interfaces with CMMS for work order and cost information, condition monitoring application for corrosion rate and RBI software for analysis results. Inspection key performance reports (such as past due inspections) can be viewed on the intranet by anyone within the company without logging into the Inspection Data Management System.

g. Reassess risk analysis. In an ideal situation, level of damage should be predictable using the risk-based tools. In the real world, there is always a variation. The only way to come closer to reality is to continuously correct the input. Any time the damage is not at the predicted level, the RBI input should be checked to determine the cause for variation. Also, input for similar service equipment should be reviewed and corrected. Long term, this process will help predict the damage accurately.

2. **Focused Reliability Improvement Programs:** Ideally, this element is not required if properly functioning inspection management exists. However, having focused reliability improvement programs to address issues common to multi sites can make a step improvement. Generally these programs require significant spending. A systematic approach should be used to "sell" the program. It is critical to collect past failure history and associated cost due to lost production, reactive maintenance, safety and environmental issues. The CUI triangle given in Attachment 5.0 is a good example of a metric to justify a CUI program. The number in the top red section is the incidents that resulted in plant outages, flaring, fires and major losses. The number in the middle yellow section is near leaks or near misses found during CUI inspections that required mechanical repairs. Examples are permanent and temporary repairs made during plant operation, repairs required during the next turnaround and pin holes leaks where a leak clamp was installed. Without the CUI program, a significant number of incidents in the middle yellow section would have moved up to the top red section resulting in plant outages or fires. The number in the bottom green section is the CUI findings not requiring mechanical repairs. These findings are those where metal loss is within the acceptable limits. These areas are blast cleaned, painted and insulated. Ideally CUI damage should be predicted and arrested in the bottom "green" section. Otherwise, these incidents will move up in the CUI triangle and will show up as incidents involving plant outages, fires or major asset loss.

   Develop a risk-based remediation plan that includes the investment amount and return on investment. It will be evident from this exercise that these programs are truly profit-generating initiatives. These are long-term initiatives, and it is critical to have continuous and sustained management support. Quite often in a large company, the reliability and inspection personnel are faced with new management personnel due to rotation. The success of these programs lies in having management and technical champions who continuously educate the stake holders and decision makers on the impact of these programs on the profitability of the company.

   At Lyondell/Equistar, the focused reliability programs include CUI remediation, heat exchanger reliability(1), furnace tubes, storage tank inspections and auto-refrigeration and brittle fracture studies[3]. Detail technical requirements and work processes are developed for executing these programs. Justification for annual spending is continuously updated and communicated to management. The company is on track to spend over $200MM on these programs over 8-10 year time frame. CUI remediation is one of the major programs. Some of the key elements of this program are [4]:

- Use of ceramic coating for personnel protection. This eliminates insulation and CUI.
- Elimination of insulation where it is not required for personnel protection or energy conservation. This could result in eliminating 20-40% of the insulation in the CUI temperature range. Remember, if there is no insulation, there can be corrosion under insulation.
- Use of rope access technique to seal insulation termination points to prevent water ingress. Also this technique is used for CUI remediation work where it is cost effective[5]
- Use of subliming type fireproofing to replace high maintenance light weight fireproofing.
- Use of coating that can be applied on damp or sweating service to reduce work during a turnaround.

At Lyondell/Equistar, "technical directives" are issued to address reliability and safety issues that are global in nature and require immediate attention. These documents are issued by the vice president of engineering or manufacturing to the plant manager, and they contain very specific action items and due dates for completion. One good example is the technical directive issued to remediate small bore piping branch connections.

3. **Failure Reporting and Investigation:** Ideally, there should not be any failures if a properly functioning inspection program exists. However, in reality, some failures are bound to occur. The saying, "If only we knew what we already know" is too true for the process industry. There are so many failures and near misses at each facility that significant improvements can be made just by learning from those incidents.

   Best-in-class companies have strong management systems for incident reporting and investigation. In the stationary equipment area, leaks and near leaks are key data for improving reliability. Every leak is an opportunity to learn why the inspections program failed to predict it. Classifying leaks into reliability, environmental and safety categories help prioritize action items and gain visibility. The level of investigation depends on the significance of the incident. Metallurgical failure analysis is required wherever the failure is not obvious. Engineered solutions and action plans are developed to prevent similar failures in the future. The critical activity is to IMPLEMENT the recommendations[6]. This is where most of the systems break down. We suggest having a central system to capture all action items and issue past-due action-item lists to management and the action owner at specified time intervals. The cycle is not complete until the equipment or system is monitored to determine if the action plan was successful. Only then is the root cause eliminated or managed within the acceptable tolerance.

Operations Discipline

A unit experienced a major leak in a column due to internal corrosion. The incident investigation indicated that a process change was documented and approved through the

Management of Change (MOC) Process. RBI analysis did not predict the leak and suggested inspection 10 years from now. What happened? The investigation further indicated that inspection personnel were not included in the MOC approval process; therefore, the RBI analysis was not updated with the higher corrosion rate expected from the change in process.

It is critical to the success of a mechanical integrity program that operations, process and maintenance personnel understand the inspection activities and work processes. At best-in-class facilities, the applicable MOC forms are routed through the inspection department for review, approval and updating of inspection planning tools. Operators are trained on mechanical integrity critical variables. When these critical variables are exceeded, inspection personnel are notified, an investigation is performed and an action plan is developed.

Some other examples where operations activities and work processes can greatly enhance stationary equipment reliability are:

- Integrating the routine visual inspection of equipment and piping into operators' rounds. Use of operators as first line inspectors greatly enhances the inspection program. An example is visual inspection of insulation for damage to jacket and caulking. This activity helps avoid CUI by maintaining insulation integrity.

- Maintaining cooling water quality and back flushing cooling water exchangers. This will have a significant positive impact on heat exchanger tube bundle reliability.

- Reporting all pressure boundary leaks irrespective of leak rate. Inspection personnel must be notified whenever there is a leak of any degree.

Implementation Discipline

An exchanger leaked within a few months after a turnaround. This exchanger was retubed during the turnaround. Tube metallurgy was upgraded from carbon steel to duplex stainless steel (SS), and tube-to-tubesheet joints were welded to prevent leaks to the shell side. An investigation indicated that the tube-to-tubesheet welds had cracked. Engineering Standards, which normally would be used for procuring new exchangers, had not been applied to this work. This omission resulted in the weld procedure for duplex SS to carbon steel weld not having been reviewed. Also, NDE requirements for the welds had not been specified.

Projects, maintenance and turnaround activities are intended to improve reliability of the plant. This is the time when a change in terms of adding new equipment, modification or replacement is made. Generally, most of the activities are related to preventive or predictive maintenance. If the engineering and quality requirements are not rigorously applied, the result could be an incident similar to the one described above. The four key elements in this focus area, when properly implemented, should prevent such incidents. They are:

1. Maintenance / Turnaround Planning and Execution
2. Project Engineering and Execution
3. Purchasing and Materials Management
4. QA / QC Programs

1. **Maintenance / Turnaround Planning and Execution:** The following steps are suggested to make a turnaround cost effective and achieve expected reliability results.

    a. Plan the inspection and maintenance activities based on RBI analysis. Significant cost savings can be realized by performing external inspections in lieu of internal inspections where the equipment is in non-corrosive service.

    b. Perform pre-turnaround inspections if there are deficiencies in the evergreen inspection program. This will minimize surprises and unplanned expenses during the turnaround.

    c. Set the technical and quality requirements in the job package for each scope item. Both engineering and inspection personnel must review and approve the package. Inspection and test plans are included in each package.

    d. Define the job closure process to include inspection personnel sign off to indicate that all work and associated inspection activities are complete.

    e. Establish performance metrics such as weld reject rate, rework rate and non-conformance rate and publish them daily during the turnaround.

    Remember, the maintenance and turnaround opportunities are meant to solve and eliminate the problem, not cause new once. Do it right the first time.

2. **Project Engineering and Execution:** Use the total-cost-of-ownership philosophy. Best-in-class companies establish design life for each equipment type and design the equipment to meet those criteria. The following steps are critical for constructing a highly reliable plant or equipment:

    a. Use consistent work practices, workflows and tools for justifying projects.

    b. Use Engineering Standards that define mandatory minimum requirements. These standards provide design, fabrication, inspection, testing and construction requirements. Management systems exist for disposition of deviations to engineering standards.

    c. Use a qualified engineer (with functional knowledge) to review and approve data sheets and specifications during the design phase. Reliability and inspection personnel provide input on applicable jobs. Front-end loading of the project is critical to minimize total project cost.

    c. Add inspection and test plans for shop and field work. A surveillance plan is in place to monitor QA and QC.

    d. Establish alliances with engineering contractors for large and small projects. Most often engineering contractors that manage large projects can not effectively execute small projects and vice versa. Working with the same contractor has benefits in their knowing the customers' work processes and technical and quality requirements.

e. Use an approved vendor list to supply equipment (both commodity and engineered) and services. A management system exists for disposition of deviations to the approved vendor list.

f. Perform shop and field surveillance to assure that the requirements are met. A management system exists for disposition of non-conformances.

g. Provide project documents to the plant for updating equipment files. The plant develops a condition monitoring plan.

3. **Purchasing and Materials Management:** The purchasing and materials departments play a key role in the reliability program. An approved vendor list for equipment (engineered and commodity) and services is critical for success. Best-in-class companies have provider qualification programs that consist of a review of safety, technical, quality and commercial capabilities. The system at Lyondell/Equistar is called Global Approved Providers System (GAPS). The system requires a rigorous vertical slice audit to verify technical and quality capabilities. Randomly a job is selected and walked through all stages - from raw material receipt to finished product - to verify implementation of all the elements of the vendor quality manual. Findings and observations are documented and reviewed with the vendor management for acceptance. The vendor is required to close the audit findings prior to being included in the company vendor list. The auditors will specify a rating, range of qualifications and restrictions which will be documented in GAPS. Purchasing personnel use this information prior to ordering equipment or services. The need for follow-up audit is determined based on vendor performance.

   It is important to have a limited number of vendors for each equipment or service type. Where spending is large, alliance with a few key vendors is suggested to reduce total purchasing cost and also ensure quality product. Alliance vendors are generally more familiar with company workflow and technical and quality requirements. Examples where Lyondell/Equistar has established alliances are heat exchanger fabrication, NDE, inspection and QA services, commodity valves, PSV testing and inspection and gaskets. The system allows end users to enter discrepancies so that feedback can be collected and analyzed for further action. Purchase requisitions have clear technical and quality requirements. Purchasing personnel are trained in company standards and quality systems to ensure that all appropriate documents are included in the request for quotation (RFQ) document.

   The warehouse also plays an important role in reliability. Receiving inspections, including positive material identification (PMI) and dimensions checks, are defined by the individuals submitting the requisition. Warehouse personnel are responsible for arranging receipt inspections, marking, storing and issuing the correct material. Major incidents could occur if the focus is just on receiving correct material rather receiving, storing and issuing correct material. Good examples are loose items such as gaskets, plugs, nipples, nuts and bolts. Proper controls in the warehouse are essential to ensure that correct materials are issued. Risk-based field PMI checks are required to verify the materials.

4. **QA/QC Programs:** QA/QC programs are critical to ensure that engineering, inspection and testing requirements are incorporated during design, fabrication and construction activities. This is a requirement per OSHA 1910.119 rules. There are four steps to a quality assurance program:

   a. Pre-qualify suppliers
   b. Specify quality assurance / control requirements
   c. Implement supplier oversight
   d. Control deviations

   Qualification of suppliers was discussed in the preceding section. Best-in-class companies have quality assurance standards. These standards assure that appropriate QA requirements are uniformly specified to contractors, sub-contractors and vendors. The standards are applied to major projects, turnarounds and minor projects. Inspection, surveillance and auditing responsibilities are defined for the company, Engineering, Procurement and Construction (EPC) contractor, equipment vendor and sub-contractor or sub-supplier. For major projects, the EPC contractor is required to develop a Quality System and Project Quality Plan. This plan consists of a Procurement Surveillance Plan for EPC purchases, a Construction Surveillance Plan for EPC subcontracting and Field Inspection/Test Plans for EPC construction.

   For engineered equipment a Supplier Surveillance Requirements (SSR) are included with the purchase order to define surveillance requirements. This document describes the hold, witness and verification points. The extent of surveillance is dependent on equipment criticality, which is determined by a multi-discipline team. Standard Inspection and Test Plans (ITP) are developed for each type of field construction. The extent of field surveillance for each subcontractor is specified.

People

The individual who performed fitness-for-service evaluation resigned and left the company. This individual provided solutions to various sites but never documented assumptions and calculations in the report. Also, lessons learned and best practices identified through various fitness-for-service evaluations were not documented. Suddenly the company found itself in a vacuum. There was no knowledgeable person available to perform the calculations. Also, it was impossible to reproduce past calculations.

When personnel rotate, retire or resign, much corporate memory leaves with the individual. Due to cost pressures, companies cut back on training and development. Succession planning for technical specialists is literally non-existent in some companies. Yet, very few companies have systems to capture individual knowledge. The following methods are used to capture knowledge at Lyondell/Equistar:

- **Standards and Guidelines:** Work processes and technical requirements that must be followed are incorporated in the standards. Engineering Guidelines are a series of documents that capture best practices identified across the company. Examples of Engineering Guidelines are API level 1 and level 2 fitness-for-service evaluations, CUI program development, inspecting for external stress corrosion cracking and heat exchanger tube testing methods.

- **Best Practice Teams and Annual Reliability Forum:** Best practice teams are formed in each discipline. The Stationary Equipment Best Practice Team is a forum to discuss and gain alignment on engineering, reliability, mechanical integrity and inspection technology issues. The team consists of inspection supervisors from worldwide sites and stationary equipment specialists from the plants and central engineering. The team shares best practices, successes and lessons learned. The team meets formally every quarter for 1-1/2 days. The team provides direction for major programs such as CUI, tank inspection program and risk based inspection.

  Worldwide Reliability Forums are conducted every year at one of the major manufacturing facilities. Approximately 250-300 engineers, specialists and management personnel from worldwide facilities attend the 3-day forum. This presents an opportunity to share and learn best practices in various disciplines.

- **Technical training:** This training is conducted by company technical specialists. The training focuses on technical requirements, work processes and applicable company standards. Training is custom-designed for the audience. Training materials are made available on the company intranet for individuals to refresh themselves at anytime.

- **Technical Advisory Series:** This series is a communication tool for distributing technical information across the company. They are one-page documents intended to create awareness of pertinent reliability and technology issues.

- **Mentoring:** This program is an excellent opportunity to train younger engineers to become discipline specialists. Younger engineers are assigned to a certain field and asked to work closely with the senior specialist. A good example is development of engineers for performing fitness-for-service evaluation.

Organization structure, job description, workflow, roles and responsibilities are very critical to the success of reliability programs. At Lyondell/Equistar, the unit inspector concept has worked well to make a step improvement in the stationary equipment inspection and reliability programs. Unit inspectors coordinate stationary equipment inspection program for a unit or area. These individuals manage the inspection program rather than simply perform inspections. They are knowledgeable in all applicable company and industry standards, RBI, inspection methods and workflow. Unit inspectors must be company employees in order have ownership, responsibility and accountability. (Field API inspections and NDE can be performed using qualified contract personnel.) There is a technical ladder for unit inspectors with three positions on the ladder (entry level, career level, advanced level). Core competencies, expectations and job description for each position are defined. Attachment 6.0 specifies core competencies used at Lyondell/Equistar and an example of behaviors expected.

Benchmarking is done to determine inspection department optimum staffing level for each site. The number of pressure vessels (drums, columns and heat exchangers) per inspector is a good measure to determine the required number of unit inspectors. The rest of the workload becomes proportional to pressure vessel count. Depending on the complexity of the unit, 225 to 300 is the optimum number of pressure vessels per inspector. Operating units should be divided among inspectors based on pressure vessel count. Attachment 7.0 provides a suggested inspection organization.

The reliability superintendent is a key position in a plant. In general, successful reliability superintendents have functional knowledge and experience in inspection management. The following are the key factors for success [6]:

- Leadership
- Credibility
- Communication skills
- Interested in this job, not the next job
- Business perspective
- Broad view
- Technical competence

The engineer development program makes the stationary equipment engineers well rounded and knowledgeable in process, maintenance, design, reliability and project engineering. Core competencies and expectations are developed for inspection supervisors and stationary equipment engineers. Corporate technical specialists are well integrated in the manufacturing organization. Corporate specialist costs are divided among plants at a very high level rather than charged out by job number. This approach helps improve dialog between the plant personnel and corporate specialist and increases utilization of specialists on value-added assignments.

The technical specialists participate in the industry technical committees and channel industry best practices to the company. Some of the industry organizations are API, American Society of Mechanical Engineers (ASME), State Boiler Board, Materials Technology Institute (MTI) and National Association for Corrosion Engineers (NACE). Engineers make presentations at conferences and publish technical papers.

A reward program exists to recognize accomplishments and celebrate successes. This recognition program exists at department, site and company wide levels. This is key for improving moral and motivation.

Management Support

The plant was assigned a new plant manager. Suddenly the reliability organization received the edict to stop spending on maintenance painting, CUI and other key reliability programs.

This is what happens if the reliability programs do not have support at all levels of management. Management support is very critical for the success of reliability programs. It should be consistent, continuous and at all levels. Management should exhibit long-term vision, leadership and commitment. To accomplish this, persistent effort is required to educate management personnel and bring visibility to reliability issues. Best-in-class organizations have strong support for reliability programs among executive management. These organizations have reliability champions or "lobbyists" in the management ranks to make a continuous case and foster a reliability culture. In these companies, reliability culture is evident among business, research and development, Information Systems and other non-manufacturing organizations.

The test for management support and commitment is during financially difficult years. Has your manufacturing vice president or plant manager refused to reduce reliability program budget during financially difficult years? Ours have.

Summary

Reliability drives excellence in safety, cost, quality and environmental performance. In world class organizations, stationary equipment reliability always leads overall reliability. This paper presents 7 focus areas for a successful stationary equipment reliability program. Success can only be realized if these elements are **IMPLEMENTED**. A very critical element among them is management support. We suggest selling the reliability programs as "profit centers" rather than "cost centers." Return on investment should be demonstrated to gain support. Attachment 1.0 provides a check list to evaluate a Stationary reliability program. Perform a gap analysis and **IMPLEMENT** the elements that are missing and you will see a step improvement in reliability.

References

1. Gamio, Carlos and Pinto, Walter, "Shell and Tube Exchanger Reliability Study," NPRA Maintenance Conference, New Orleans, Louisiana, May 1999.
2. Hardy, Allison, Qualification Program for Heat Exchanger Tube Test Operators, NPRA Maintenance Conference, Austin, Texas, May 2000.
3. King, Ralph and Kelly, Thomas, "Auto-Refrigeration/Brittle Fracture Analysis of an Existing Olefins Plant – Lessons Learned," NPRA Maintenance Conference, San Antonio, Texas, May 2002.
4. Sanders, Joe Don, "Effectively Using Risk-Based Inspection Results to Implement a Corrosion Under Insulation Program," NPRA Maintenance Conference, New Orleans, Louisiana, May 2001.
5. Klein, Paul, "The Use of Rope Access Techniques to Inspect for Corrosion Under Insulation on Towers in an Operating Process Plant," NPRA Maintenance Conference, San Antonio, Texas, May 2002.
6. Hoffman, M. Rick, "Back to Basics, A Down to Earth Approach to Sell, Implement and Sustain Your Reliability Program," 8th International Process Plant Reliability Conference, Houston, Texas, October, 1999.

BIOGRAPHY

Walter Pinto is regional reliability manager for Lyondell Chemical Company northern region operation. He has reliability and maintenance responsibilities for multiple facilities. Previously Walter was manager of stationary equipment engineering and had worldwide responsibility for corporate stationary equipment engineering. He has 20 years of industry experience in equipment design, fabrication, inspection, reliability, maintenance and troubleshooting. He holds a MS degree in mechanical engineering from the University of Wisconsin, Milwaukee.

Attachment 1.0

MODEL STATIONARY EQUIPMENT RELIABILITY PROGRAM

Management Systems

1. The company has reliability principles and beliefs.

2. The company has standards identifying required management systems. These standards are approved by an executive officer of the company. The standards include management systems for:

    a. Developing, reviewing and approving standards and procedures.
    b. Approving addenda and deviations to the standards and procedures.
    c. Establishing inspection priority and dates using RBI methodology.
    d. Developing, approving, implementing and updating inspection plans.
    e. Timely notifying Site Management of upcoming inspections and recommendations, past due inspections and recommendations, and deficiency.
    f. Performing condition assessment on all equipment inspection results/data in a timely manner.
    g. Performing fitness-for-service evaluation to resolve deficient stationary equipment conditions found during condition assessments.
    h. Prioritizing, reviewing, tracking, and closing inspection recommendations and corrective actions.
    i. Approving inspections deferrals and recommendations. Deferral form contains reasons for not executing action, consequences, resolution plan including a new date for completing the action and technical justification for delaying inspections and recommendations. Plant Manager approves this request.
    j. Conforming to standards and specifications for maintenance materials, maintenance repairs, alterations, and new stationary equipment procurement and construction.
    k. Documenting non-conformances, their resolution and approval of the resolution plan.
    l. Reporting and analyzing inspection key performance indicators (KPI) and to take improvement actions.
    m. Ensuring equipment history and inspection records are maintained.

3. Each site has plant procedures defining requirements for inspections and maintenance activities. The procedures include roles and responsibilities and workflow. Suggested inspection and maintenance procedures are given in Attachment 3.0.

4. The company has inspection and maintenance standards for defining technical requirements related to stationary equipment maintenance.

5. Each inspection and maintenance standard and procedure has a single owner who is held responsible and accountable for its content.

6. Addenda and deviations to the standards and procedures are approved by the standards owner.

7. Performance-related actionable key performance indicators are established to improve reliability and cost objectives. As a minimum, the following KPIs are reported:

   a. Past due inspections: Inspections that have not been performed in accordance with the original equipment inspection plan.
   b. Past due recommendations: Recommendations that have not been completed by the original due date.
   c. Relief valve failing pretest: Separate data for number of valves failing to open >110% of set pressure and number of valves lifted <90% of set pressure.
   d. Number of Leaks: Number of pressure boundary leaks.
   e. Weld reject rate: Butt weld radiography pass/fail percentage.
   f. Percentage of parts Rejected: Percentage of parts inspected that did not meet the defined criteria.

8. KPIs are analyzed and action plans are implemented.

9. Internal and external bench marking is done to evaluate staffing and cost structure.

10. Audits are performed at a fixed interval to verify performance of management systems and programs.

11. A detail audit protocol exists.

12. The auditors are trained in the audit protocol and are knowledgeable in the company's performance, inspection and maintenance standards.

13. Audit findings and observations are documented and reviewed with the plant manager and his/her staff at the end of the audit. The action items are monitored by management. Closure is required within a fixed time frame.

Engineering Practices

14. The company has global Engineering Standards defining technical requirements for design, construction, inspection, testing, erection and commissioning of new equipment and plants.

15. The company has global piping and valve specifications.

16. Each Engineering Standard has a single owner who is held responsible and accountable for its content. The piping and valve specifications have single ownership.

17. Use of Engineering Standards is mandatory for all new and existing equipment. This has been endorsed by the executive management.

18. Plant and project specific technical requirements are captured in addenda to the Engineering Standards. These addenda are reviewed and approved by the original standard owner.

19. A deviation to an Engineering Standard is approved by the owner of the standard.

20. Engineering, Inspection and Maintenance standards and procedures are administered by a dedicated group.

21. All standards and procedures are available on the company intranet for easy access.

22. A system exists for receiving and disposition of improvements suggested to the standards. Lessons learned from projects and turnarounds are promptly implemented by the standard owner.

23. An accurate list of stationary equipment including pressure relief devices and piping circuits/line numbers exists.

24. Equipment purchase specification data are stored in a query-able format in a master database. Duplicate databases are strictly prohibited.

25. Design calculations, drawings, vendor documentation and regulatory documents are stored in a central master system.

26. P&IDs, PFDs, equipment lists, equipment drawings and datasheets are updated promptly when modifications or repairs are made to the equipment.

27. Computer-Based Maintenance Management System (CMMS) exists to document equipment maintenance work history, failure history and cost.

Reliability Practices

28. Each site has a corrosion manual. It is developed with the input of process, inspection and metallurgical engineers. The corrosion manual contains:

    a. Process description for each unit and section.
    b. Corrosion table listing product characteristics, process conditions, corrosion mechanism and failure history.
    c. Failure history at similar process plants within the company and in the industry.
    d. Information on current metallurgy and improvements made to the metallurgy over the plant life.
    e. Key corrosion variables and operating envelopes.

29. Operators and engineers are trained in the content of the corrosion manual.

30. Investigation is performed when operating conditions deviate from mechanical integrity critical variables and the operating envelope.

31. Quantitative and/or qualitative risk analysis is performed for prioritizing equipment for inspection and developing inspection plans. As a minimum, the following risk based tools are available:

    a. Pressure vessels, heat exchangers and piping circuits (such as RBI)
    b. Heat exchanger bundles
    c. Pressure relieving devices
    d. Storage tanks

32. Unit inspection plans are developed in the second quarter for next calendar year. This plan is used to develop inspection and associate maintenance activity budget for the unit.

33. 90-day look ahead report indicating upcoming inspections is published each month to concerned plant personnel.

34. Company and contract inspection and non-destructive examination personnel are qualified to the industry-accepted standards.

35. A core group of contract non-destructive examination personnel are identified for ultrasonic shear wave inspection, heat exchanger tube testing, acoustic emission testing and other high technology inspection methods. An in-house qualification program exists to qualify tube-testing technicians.

36. A computer-based system exists to document and store condition-monitoring data such as thickness data (examples, UltraPipe, Meridium Condition Manager).

37. A computer-based system exists to document the following:

    a. Data used for risk analysis and inspection priority
    b. Equipment specific data
    c. Unit inspection plans and specific equipment inspection plans
    d. Inspection due date
    e. Inspection results and reports
    f. Findings and observations
    g. Condition assessment and fitness-for-service evaluation reports
    h. Recommendations prioritization and closure
    i. Work order information related to recommendations

38. Inspections are performed on a timely basis in accordance with the due date. Findings and observations are documented in the inspection report.

39. Evaluation of inspection results (condition assessment) is performed to determine if findings and observations meet the acceptance criteria.

40. Corrective actions are developed to remediate equipment deficiency and tracked to closure. A system exists to prioritize recommendations.

41. Inspection results are used to reevaluate RBI risk and determine next inspection date.

42. Focused reliability programs exist for issues common to multi-units or sites and situations requiring intensive resources. Examples: Corrosion-under-insulation remediation and heat exchanger reliability programs.

43. A mechanism exists for senior management to direct plants to prioritize and address specific reliability or safety issues that are global in nature and require immediate and sustained attention.

44. Detail technical requirements and work processes are developed for executing focused reliability improvement programs.

45. All pressure boundary leaks, irrespective of how small, are being reported as incidents.

46. Metallurgical analysis is performed where cause of the failure is not known.

47. Engineered solutions and action plans are developed to prevent similar failures in the future.

48. Solutions are applied to similar service equipment or systems.

49. System exists to capture and track all action items to closure.

50. A past due action item list is issued to management at a fixed time interval.

51. The failed equipment or system is monitored to determine if the action plan was successful.

Operations Discipline

52. Operators perform and document routine visual inspection of equipment and piping. Deficiencies are identified and communicated to inspection organization.

53. Operators are trained on mechanical integrity critical variables and operating envelope.

54. Investigations are performed and action plans are developed when mechanical integrity critical variables are exceeded.

55. A management of change (MOC) process exists to document and approve changes to process chemicals, technology, equipment and procedures.

56. Applicable MOC forms are routed to the inspection department for review, approval and updating inspection planning tools.

Implementation Discipline

57. An evergreen turnaround scope item log is maintained to document required inspections and maintenance work as a result of inspection recommendations. Risk based tools are used for identifying inspection and maintenance scope items.

58. Pre-turnaround inspections are performed where deficiency exists in the evergreen inspection program.

59. A job package is developed for each scope item. Technical and quality (Inspection and Test Plans) requirements are included.

60. Engineering and inspection personnel review and approve the job packages.

61. The job closure process requires inspection personnel sign off to indicate that all work and associated inspection activities are complete.

62. Performance metrics such as weld reject rate, rework rate and non-conformance rate are established and published.

63. Total cost of ownership philosophy is used to justify projects. Consistent work practices, workflows and tools are used.

64. Engineering standards that define mandatory minimum requirements are used. Management systems exist for disposition of deviations to engineering standards.

65. Data sheets and specifications are reviewed and approved by a qualified engineer during the design phase. Reliability and inspection personnel provide input on applicable jobs.

66. Inspection and test plans are added for shop and field work. A surveillance plan is in place to monitor QA/QC.

67. Alliances agreements are established with engineering contractors for large and small projects.

68. An approved vendor list is used to procure equipment (commodity and engineered) and services. A management system exists for disposition of deviations to the approved vendor list.

69. Shop and field surveillance is performed to assure that the requirements are met. A management system exists for disposition of non-conformances.

70. Project documents are turned over to the plant in a timely manner for updating equipment files and developing condition monitoring plan.

71. The approved vendor list is maintained for supplying equipment (commodity and engineered) and services.

72. The provider qualification program exists to qualify vendors. This program reviews safety, technical, quality and commercial capabilities. Audits are performed to verify vendor's technical and quality programs.

73. Alliance agreements exist with critical suppliers.

74. A system exists to collect feedback on the current vendors. Non-conformances are analyzed and actions taken.

75. Purchasing personnel are trained in company standards and quality systems.

76. Purchase requisitions contain technical and quality requirements. Receiving inspection requirements including positive material identification (PMI) and dimensions checks are specified.

77. A management system exists for performing receipt inspections, marking, storing and issuing materials.

78. Quality assurance standards exist for defining QA requirements for contractors, sub-contractors and vendors.

79. Inspection, surveillance and auditing responsibilities are defined for the owner company, EPC (engineering, procurement and construction) contractor, equipment vendor and sub-contractor or sub-supplier.

80. Supplier Surveillance Requirements (SSR) is included with the equipment purchase order.

81. Inspection and test plans are developed for field work.

82. Non conformances are reported and corrective actions to rectify non conformance are taken.

People

83. A system exists to document and publish best practices, lessons learned and successes.

84. Best practice teams exist to gain alignment on engineering, reliability, mechanical integrity and inspection technology issues.

85. Technical training is conducted to educate and train engineers, inspectors, operators and technicians.

86. Communication tools exist for distributing technical information and creating awareness of pertinent reliability and technology issues.

87. A mentoring program exists.

88. Unit inspectors manage the inspection program for their unit.

89. A technical ladder exists for unit inspectors.

90. Core competencies and expectations are developed for each position.

91. Benchmarking is done to determine the optimum number of inspectors, NDE technicians and clerical support. The optimum number of pressure vessel count per inspector is in the range of 225 to 300.

92. A dedicated inspection supervisor exists for sites with more than 600 pressure vessels.

93. A development and rotation program exists for engineers. Core competencies and expectations are developed for inspection supervisors and stationary equipment engineers.

94. Corporate specialist costs are divided among plants at a very high level rather than charged out by job number.

95. Technical specialists participate in the industry technical committees and they channel industry best practices back to the company. They make presentations at conferences and publish technical papers.

96. A reward program exists at multiple levels of the organization to recognize accomplishments and celebrate success.

Management Support

97. Management has long-term vision and is committed to investing in reliability improvement programs. Support and commitment is obvious during financially difficult years.

98. Strong support exists for reliability programs among executive management.

99. Return on investment is calculated to support investment on reliability improvement programs.

100. Reliability champions or "lobbyists" exists in the management ranks to foster a reliability culture.

Attachment 2.0: Reliability Related to Safety and Cost Performance

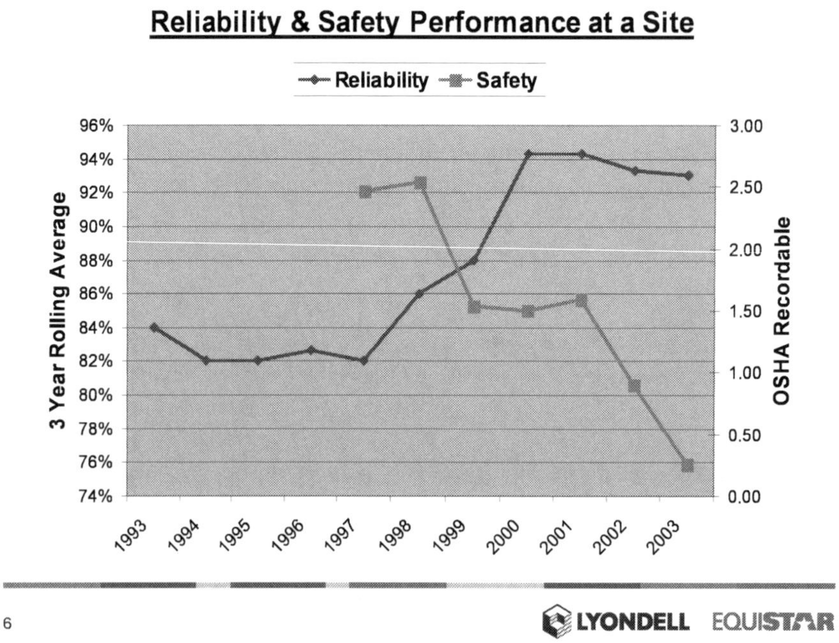

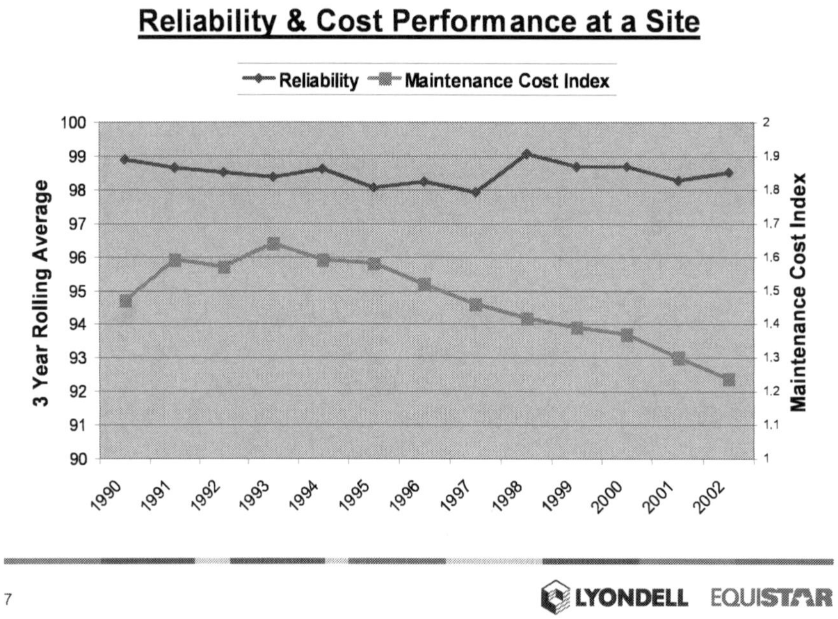

Attachment 3.0: Suggested inspection and maintenance procedures

Procedure Title
Administrative Procedures
Organization and Responsibilities
Management of Inspection Procedures
Inspection Personnel Qualifications
Past Due Inspection and Recommendation Notification
Management of Recommendations
Inspection Records and Security Management
Reporting and Analyzing Inspection Key Performance Indicators
Management of Non-Conformance
Inspection Procedures
Equipment Classifications
Unit and Equipment Inspection Plans
Equipment Condition Assessment
Mechanical Integrity Risk Assessment (MIRA)
Equipment Failure Investigation
Receipt Materials Quality Control
Equipment Procurement QA and Construction QC
Heat Exchanger Tube Inspection, Retubing and Plugging
RV Inspection and Repair
Turnaround Inspection and Repair Scope Development and Execution
Inspection Work Instructions
Pressure Vessel Inspections
Heat Exchanger Bundle Inspections
Piping Inspections
Dynamic Equipment Inspections
Atmospheric and Low Pressure Storage Tank Inspections
Fired Boilers and Heaters Inspections
Structures and Building Inspections
Radiographic Film Interpretation
Ultrasonic Thickness Measurement
Liquid Dye Penetrant Inspection
Magnetic Particle Inspection
Eddy Current Inspection
Hydrostatic and Pneumatic Testing
Maintenance Procedures
Hot-taping and Stoppling
Welding Repairs, Alteration and Rerate of Stationary Equipment
Flange Bolt-up Procedures
Temporary Repairs (leak clamps)
Maintenance of Coating, Insulation Systems, and Fireproofing

Attachment 4.0:

Leak Reduction Success

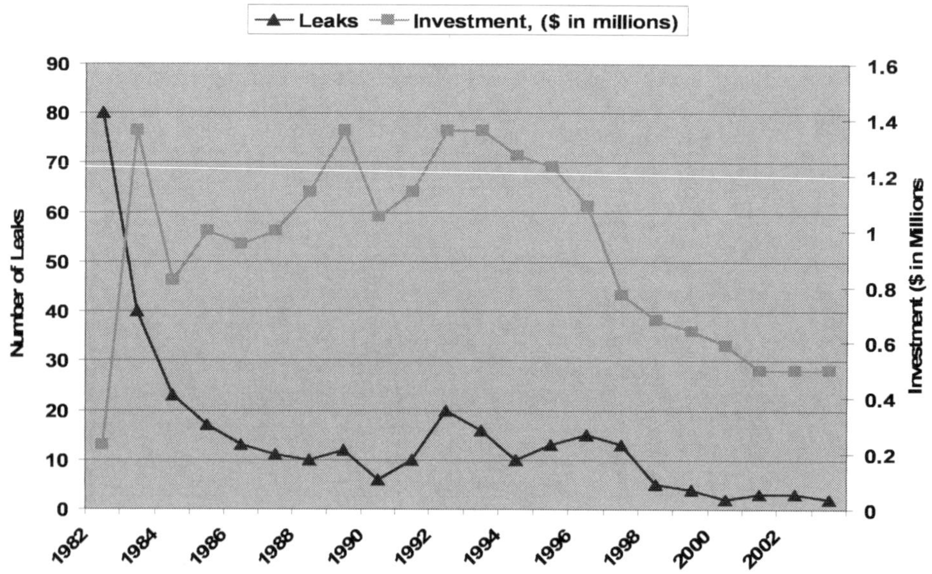

58

LYONDELL EQUISTAR

Attachment 5.0: CUI Triangle

CORROSION-UNDER-INSULATION FINDINGS

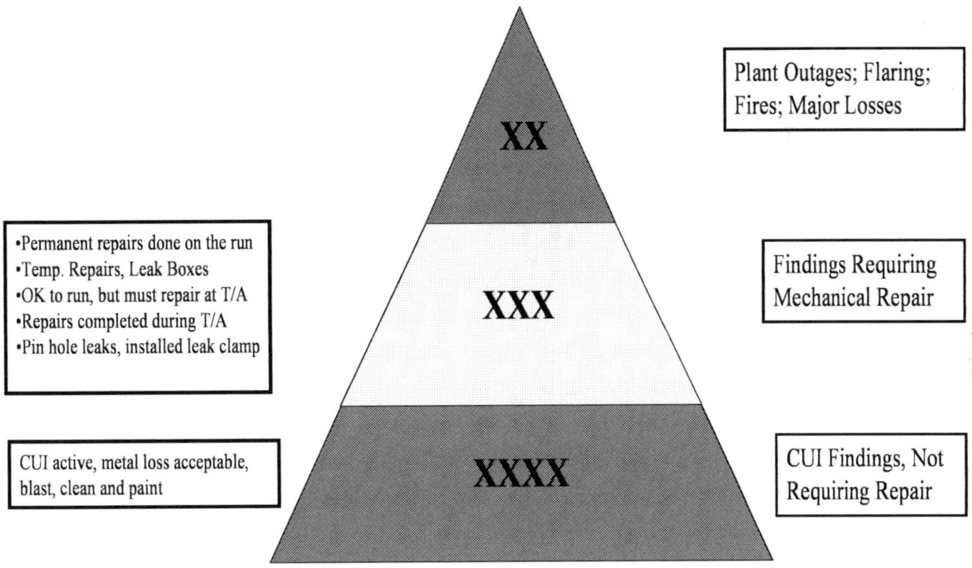

33 LYONDELL EQUISTAR

Attachment 6.0: Inspector Core Competencies

The following core competencies are used for inspectors at Lyondell/Equistar. Required qualification for each position is defined in the job description.

1. Technical Knowledge
2. Industry Codes and Standards
3. Inspection Workflow Process, Internal Standards and Procedures
4. Equipment Classification (RBI)
5. Routine Inspection Planning and Execution
6. Turnaround/Shutdown Inspection Planning and Execution
7. Inspection Personnel Qualification
8. Evaluation of Inspection Information
9. Inspections Recommendations Management
10. Maintenance QA/QC
11. Failure Analysis, KPI and Evergreening Inspection Plans
12. Inspection Records Management
13. Reliability Improvement Programs
14. Business Ethics and Personnel Standards
15. Communication Skills and Customer Orientation
16. Leadership Skills, Team Work, Initiative and Develop Others
17. Inspection Management Tools

Example of expectations for "Equipment Classification" competency:

Competencies	**Expectations** (Note 1)		
	Inspector	**Senior Inspector**	**Principal Inspector**
Equipment Classification	• Independently performs qualitative (Level I) RBI analysis to determine total risk, next inspection date and type of inspection. • Demonstrates ability to perform quantitative (level III) RBI analysis ***under supervision***. • Determines priorities for heat exchanger bundle inspections using Enterprise tools (e.g. RAT) under supervision. • Determines RV criticality ***under supervision.*** Determines inspection interval for State/Local regulatory body controlled equipment. • Reviews management of change documents for impact on RBI analysis.	• ***Independently*** performs ***quantitative*** (Level III) RBI analysis to determine total risk, next inspection date and type of inspection. • Demonstrates understanding of API 580 and 581 requirements. • Independently determines priorities for heat exchanger bundle inspections using Enterprise tools (e.g RAT). • Proactively ***optimizes*** RV testing interval using past inspection/testing data. • Interacts with operations for information impacting damage mechanism and RBI analysis.	• Demonstrates ***thorough*** knowledge of likelihood and consequence analysis per API 580 and 581. • Demonstrates ***thorough*** knowledge of the logic in software tools for inspection (e.g. RAT, API-RBI). Understands sensitivity of the output on each input field in the software. • ***Proactively*** initiates discussions with Reliability and Process Engineering on alternate designs to meet optimum RV testing interval. • Considered an ***expert*** in Enterprise qualitative tools for inspection (e.g. API-RBI, RAT) • Proactively establishes systems to ensure timely input from operations when operating variables impacting equipment damage exceeds the acceptable limits.

Note 1: Expectations are cumulative. Individuals must demonstrate all expectations at their grade level and all levels below.

Attachment 7.0: Suggested Inspection Organization Structure

Best in Class Inspection Organization Structure

Pressure vessels per unit inspector : 220

Note: All four unit inspectors are company employees

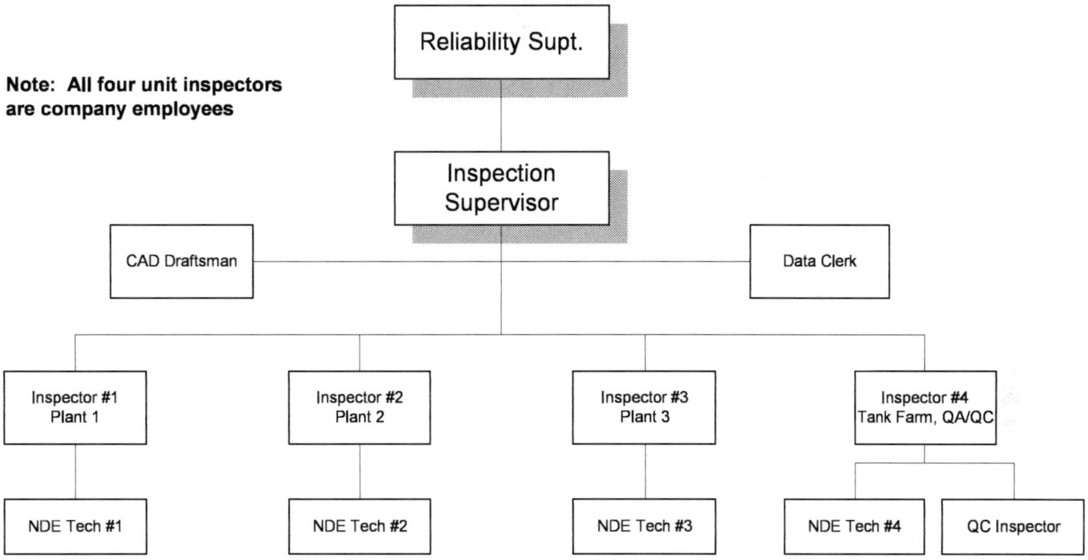

Inspector #1
- 280 pressure vessels
- 504 piping circuits
- 254 PSVs

Inspector #2
- 329 pressure vessels
- 592 piping circuits
- 332 PSV's

Inspector #3
- 186 pressure vessels
- 335 piping circuits
- 200 PSV's

Inspector #4
- 83 pressure vessels
- 149 piping circuits
- 78 ASTs
- 186 PSV's

Attachment 8.0: List of Abbreviations

API	American Petroleum Institute
ASME	American Society of Mechanical Engineers
ASNT	American Society of Nondestructive Testing
CMMS	Computer Based Maintenance Management System
CUI	Corrosion-under-Insulation
EPC	Engineering, Procurement and Construction
GAPS	Global Approved Procurement System
ITP	Inspection and Test Plans
KPI	Key Performance Indicators
MIRA	Mechanical Integrity Risk Assessment
MOC	Management of Change
MTI	Materials Technology Institute
NACE	National Association for Corrosion Engineers
NDE	Nondestructive Examination
OSHA	Occupational Safety and Health Administration
P&ID	Process and Instrumentation Diagram
PFD	Process Flow Diagram
PMI	Positive Material Identification
PSM	Process Safety Management
QA	Quality Assurance
QC	Quality Control
RAT	Retube Analysis Tool
RBI	Risk Based Inspection
RFQ	Request for Quotation
SS	Stainless Steel
SSR	Supplier Surveillance Requirement

From PSM to Operational Excellence: Safety Critical Variables and Equipment

Tim Salmon
Lyondell Chemical Company
LaPorte, Texas
Tim.Salmon@lyondell.com

ABSTRACT

This presentation will trace the history of process safety management from prior to the implementation of the OSHA Process Safety Management (PSM) regulation to the current status of the safety critical variable and safety critical equipment programs in Lyondell. Safety critical variables and safety critical equipment are risk-based management systems that allow for extraordinary operations and reliability management vigilance.

Process Safety and Reliability management improvement is an ongoing effort in the Petrochemical Industry. Thousands of people have dedicated thousands of hours and sometimes their entire careers to ensuring we recognize and manage the hazards that exist in our plant sites. Lyondell Chemical Company has been constantly improving our process safety management systems. This paper reviews the history of process safety management in our company and the current management system that is in place in Lyondell to manage the highest level of process safety consequences. The process variables and equipment associated with monitoring and controlling the highest severity of process safety consequences are identified as Safety Critical.

Prior to 1990, the United States petrochemical industry, including Lyondell, enjoyed rapid growth with many new technologies being built in many plant sites from multiple engineering firms. For our company, most of the technology was purchased and some was developed in-house. All have since been modified from the original design. During this time, the companies designing, building, and operating the plants used common methods to review the systems. Based on these reviews, the hazards were identified and controls were developed for those hazards that could not be designed out of the process. There were variations in the philosophy and hardware of the safety system designs because of the number of different companies involved.

Throughout our plant process safety systems designs, consideration was given to the hazard initiating frequency and the severity of consequences. The hazards were addressed with both hardware and administrative systems. The methods of dealing with the various frequencies of event and the severity of consequence were not entirely consistent between the different technologies represented in our company. Industry did not have a consistent applied practice for this work. There was room for improvement, and industry as a whole continued making progress on mitigating hazards. Most notably, the introduction of preventative maintenance systems, improved electronic monitoring and control, and the introduction of more complex and flexible computing systems greatly improved plant monitoring integrity and process safety. At the same time, industry was gathering data on the integrity of equipment used on these systems.

In the 1980's as plants built in the 1960's and 1970's began to mature, it became clear to the industry that there was room for process safety improvement. Many safety system components and other equipment reached the end of their expected life, the personnel rotated to other assignments or retired, understanding of the technology and its documentation began to deteriorate, and the existing management system could not compensate for these stresses. This adversely affected the safe operation of the plants. The plants were larger and more complex, and consequently, there were more frequent and larger process safety incidents. Other companies had significant process safety incidents, including Bhopal in 1984 and Pasadena in 1989. The defining moment for our company came in July 1990 with the explosion of a 900,000 gallon wastewater tank that killed 17 at our Channelview, Texas plant site.

The response from our company was to understand the causes of the incident, and put systems in place to effectively identify and manage hazards within the processes in our plant.

Company representatives worked with OSHA to help develop PSM regulations while a comprehensive management system was developed for our company. The result of the company's investigation and response to the Channelview fatalities is our company's Operational Excellence management system. The identification and management of Safety Critical Variables and Equipment are a part of the Operational Excellence Management system.

Lyondell's Operational Excellence is an evergreen comprehensive management system that drives continuous improvement and defines the roles, responsibilities, and accountabilities for maintaining the operation of the manufacturing facilities. The requirements of our Operational Management System go well beyond the requirements of OSHA PSM. One key focus area of Lyondell's Operational Excellence program is the management of our high-consequence events to an acceptable frequency. Process variables and equipment associated with a high-consequence are labeled 'Safety Critical'. The following are Lyondell's current working definitions for Safety Critical Variables and Safety Critical Equipment.

Safety Critical Risk is defined as having the following minimum consequences and occurs at an unacceptable frequency:

- Admission of employee(s) or contractor(s) to a hospital for an overnight stay.
- Fire, explosion or hazardous material release that requires actual fire fighting or other assistance from organizations outside the facility (mutual aid).
- Release of materials that pose a threat to the environment or the health and safety of members of the public.

Safety Critical Variable: A process variable where operation outside the Safety Critical Limit would present a Safety Critical Risk.

Safety Critical Equipment: An electrical, instrumental or mechanical device designed to mitigate a Safety Critical Risk. Safety Critical Equipment includes all equipment needed to ensure monitoring of Safety Critical Variables, the display of safety critical variables, and all equipment required to function to mitigate a Safety Critical Risk.

Figure 1 shows the acceptable risk matrix for Lyondell. This figure shows increasing consequence on the x axis and increasing frequency on the y axis. The Safety Critical Risk items fall into the category IV and V on the consequence axis. We now use this consequence – frequency chart to help determine priority of activities within the company. Category IV consequences are the minimum severity for an event to be considered Safety Critical.

Category V consequences are also Safety Critical and are more severe in nature. These items require more control to ensure the required reduced frequency of event.

Figure 1
Frequency vs. Consequence

FREQUENCY YR		CONSEQUENCE I	CONSEQUENCE II	CONSEQUENCE III	Safety Critical Category
	10^{-0}	Optional	Optional	Action 2	
	10^{-1}	Optional	Optional	Optional	
	10^{-2}	No Further Action	Optional	Optional	
	10^{-3}	No Further Action	No Further Action	Optional	
	10^{-4}	No Further Action	No Further Action	No Further Action	
	10^{-5}	No Further Action	No Further Action	No Further Action	
	10^{-6}	No Further Action	No Further Action	No Further Action	
	10^{-7}	No Further Action	No Further Action	No Further Action	

Since 1990, Lyondell developed a method to determine and document the severity of consequence and to determine the frequency of initiation of the event that would lead to that consequence. Usually this determination is completed during regularly scheduled hazard reviews, but can be done anytime, during special reviews using a similar process. We use our experience and the experience of industry to determine the most likely consequence of an event. Standard industry data is used to calculate the average frequency of initiation of an event.

Once the basic review is completed, the consequence and frequency are mapped onto Figure 1 to determine if additional safeguards are needed. If needed, safeguards are added until the average frequency falls into the optional range, and if it is cost effective, additional layers of protection may be added to further reduce the frequency of the event with Safety Critical consequences.

The second key to successfully managing these high-risk consequence events is a rigorous management system that includes methods to ensure the systems are in place and effective, remain fully effective, and are evergreen. The responsibilities for managing the overall process are distributed through the corporate and site organizations. For safety Critical Variables and Safety Critical Equipment, personnel in corporate support groups, plant managers, and the plant site technical, operations, maintenance, and reliability organizations all play a key role.

Safety Critical Variables can occur in many parts of a process and are often associated with large reactor systems. One example of a Safety Critical Variable could be the oxygen concentration in an oxidation reactor gas stream. Another example could be the reactor outlet temperature on a reactor with an exothermic reaction. Each would have to be evaluated on a case by case basis. Safety Critical Equipment would be any piece of equipment needed to view the Safety Critical Variable and mitigate the hazards. Safety critical equipment tends to be interlock components and relief valves.

The Lyondell Operational Excellence management system requires the identification of the Safety Critical Equipment and Safety Critical Variables for each process. In addition, Operational Excellence standards require that we determine the current initiating event frequency, the number and type of independent protection layers, and if necessary, redesign the hardware to ensure we meet the availability frequency requirements. This only addresses the hardware and software configurations. There are many other aspects that must be in place to ensure the system is effective and remains effective that include roles and responsibilities in the plant operations, technical, maintenance, and reliability departments.

OE standards have a number of special requirements governing our work with Safety Critical Variables and Safety Critical Equipment. Safety Critical initiating events, the consequences, and the predetermined actions required to mitigate the consequences are documented in the technical files and in the operating manuals. Additional training is conducted for operating and maintenance technicians that work with Safety Critical Variables, and they are expected to take immediate, predetermined actions, including full-unit shutdowns without having to seek additional authority.

Monitoring and reporting of variances are also required. The standards require the Safety Critical variables to be monitored and compliance reports are issued daily. Operating outside our Safety Critical limits must be reported to our incident reporting system, reported across the company, and investigated at the appropriate level.

Equipment that is designed to monitor and mitigate a Safety Critical event is clearly identified in our equipment files. The reliability of the Safety Critical Equipment must be maintained to a minimum standard to ensure it is available when needed. In some cases, preventative maintenance testing, via in-service function testing, is performed to ensure the system will perform when needed. When there is a less rigorous demand on the equipment, testing is completed during outages. Deviations from a preset schedule are reported to the department managers and the site manager. Personnel maintaining the Safety Critical Equipment must also have special training to ensure their expertise is up to date.

There are other requirements to ensure the Safety Critical Equipment is in place and working as intended. When the equipment is taken out of service to perform maintenance, special precautions and notifications are required. Alternate means of protection must be identified and implemented. A review and approval process is utilized to ensure good communication and notification. Additional notifications and approvals are required if the equipment is out of service for an extended period of time. Systems that are disabled as a part of other equipment maintenance, for example, a thermocouple removed to perform maintenance, must be verified and documented as installed and functioning prior to restart of the equipment.

Compliance with site and corporate procedures must be maintained to ensure the Safety Critical management process effectiveness does not degrade over time and with personnel changes. Our corporate management system requires the sites to periodically audit themselves to check compliance with site and corporate Safety Critical standards. In addition, there are periodic evaluations completed by personnel from other sites using detailed pre-established protocols to ensure consistency. The external audits also facilitate sharing of best practices across the company.

Gaps that are discovered during both the internal and external reviews are evaluated against potential impact, prioritized, and tracked to closure in our company wide incident data base. The Safety Critical management system requires good visibility of the system's effectiveness. Items that affect the ability of the Safety Critical system to function as intended are documented and reported, and discussed quarterly with the VP of Manufacturing.

The Safety Critical Variables and Safety Critical Equipment management system in Lyondell is a proven, thorough, reliable, maintainable method to improve process safety and reliability in the Chemical Process Industry. It provides the consistent and cost-effective tools to ensure evergreen management of the most hazardous elements in our plants.

Field-based Evaluation of *PFD* Values for Safety-related Loop Typicals

Daniel Düpont, Lothar Litz

Institute of Automatic Control, University of Kaiserslautern
Kaiserslautern, Germany
duepont@eit.uni-kl.de, litz@eit.uni-kl.de

Pirmin Netter

Infraserv Höchst
Frankfurt on the Main, Germany
pirmin.netter@infraserv.com

ABSTRACT

According to IEC 61511, each safety instrumented function (SIF) in chemical production plants has to be assigned a safety integrity level (SIL). Thereby, for every SIF realization, i.e. safety-related loop, a SIL-specific probability of failure on demand (PFD) must be proven. Usually, the PFD calculation is performed based on the failure rates of each loop component, aided by commercial software tools. But this bottom-up SIL proof suffers from many uncertainties. Especially a lack of reliable failure rate data causes many problems. Reference data acquired in different environments are available to solve this situation. However, this pragmatism leads to a PFD bandwidth.

In order to make a decision for a single PFD value appropriate for plant applications in chemical industry, a data ascertainment has been initiated by the European NAMUR within its member companies in 2002. Since 2003 more than 12,000 single-channel and 10,000 dual-channel safety-related loops have been observed in detail. Combined with statistical methods their results display large deficiencies of the bottom-up approach.

1. INTRODUCTION

Especially in chemical and pharmaceutical industry, plant safety plays an important role. To guarantee a homogenously high safety standard around the world a global directive was created by IEC 61511 [1]. The compliance with this guideline imposes strict requirements to plant operators. One central term of IEC 61511 is given by the safety instrumented system (SIS) which executes one or more safety instrumented functions (SIFs). If the process tends to enter a dangerous range, the SIS has to interfere in order to prevent any harm to humans and the environment.

For each newly-installed SIF, IEC 61511 prescribes to perform a SIL assessment, see Fig. 1.

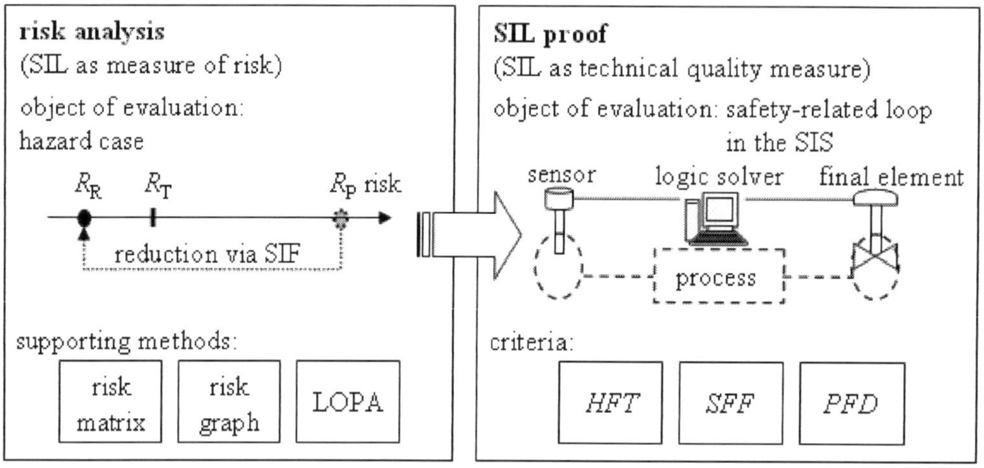

Fig. 1 – SIL assessment

To determine the process risk R_P for a certain hazard case, a risk analysis must be performed. If R_P lies above the tolerable risk R_T, all supporting methods result in one of the four safety integrity levels (SIL 1 to 4) as measure of risk. To reduce R_P of each scenario with SIL classification, a SIF must be identified and installed in the SIS. Thereby, a risk reduction to a residual risk R_R below the tolerable one must be created.

The SIF realization as hardware loop in the SIS has to fulfill SIL-specific criteria. Its SIL conformity must be proven under consideration of several, predominantly quantitative constraints. These quantities describe the structural and technical quality of the loop. A key criterion creating a SIL is given by the probability of failure on demand (*PFD*). In a strict mathematical sense the *PFD* is the conditional probability P(F/D) describing the probability of a SIF failure (F) under the condition of a SIF demand (D). From the technical point of view the *PFD* is given by the SIF unavailability which can be calculated in two different ways: Bottom-up methods are appropriate to determine the *PFD* of any system consisting of components with known data (failure rates) [2]. Usually, they are used for the SIL proof. Top-down methods on the other hand originate from real loops and are based on their statistics. The ideal case would be represented by similar results for the outcomes of both, the bottom-up and the top-down approach. Applying these methods and comparing their results can answer the interesting question:

Is the often used bottom-up SIL proof able to map realistic *PFD* values?

2. NOMENCLATURE

ΔT	=	Observation period
$1\text{-}\alpha$	=	Confidence level
BPCS	=	Basic process control system
FIT	=	Failures in time, i.e. per 10^9 hours
k	=	Number of occurred dangerous, undetected failures
$MooN$	=	M out of N voting
MDT_{DU}	=	Mean down time by dangerous, undetected failures
MDT_{DUrel}	=	Relative mean down time by dangerous, undetected failures
$MTTD_{\mathrm{DU}}$	=	Mean time to detection of dangerous, undetected failure
$MTTR$	=	Mean time to repair
n	=	Number of individuals (loops)
$p(\Delta T)$	=	Probability of dangerous, undetected failures related to ΔT
PFD	=	Average probability of failure on demand
PFD_{MooN}	=	PFD of a $MooN$ system
PLC	=	Programmable logic controller
$p_{\mathrm{low}}(\Delta T)$	=	Lower $p(\Delta T)$ confidence interval boundary
$p_{\mathrm{up}}(\Delta T)$	=	Upper $p(\Delta T)$ confidence interval boundary
R_{P}	=	Process risk
R_{R}	=	Residual risk
R_{T}	=	Tolerable risk
SIF	=	Safety instrumented function
SIL	=	Safety integrity level
SIS	=	Safety instrumented system
T_{I}	=	Proof test interval
β	=	Common cause failure factor
λ_{DD}	=	Rate of dangerous, detected failures
λ_{DU}	=	Rate of dangerous, undetected failures
λ_{S}	=	Rate of safe failures

3. PFD CALCULATION ACCORDING TO IEC 61508

3.1 General Approach (Bottom-Up)

In order to get representative results, special SIF realizations have to be defined which can serve as a kind of benchmark. Afterwards, reliability data are collected for each loop component. Finally, these so-called typicals serve as basis of the loop evaluation in commercial software tools.

Gathering component reliability data is the most critical step of *PFD* determination. Component failure rates as basis of calculation strongly influence the *PFD* height and quality. Normally, they should be provided in SIL conformity declarations given by the manufacturers. Concerning the availability and the content of these documents, several cases are to be distinguished:

case 1: A manufacturer`s declaration is available containing trustworthy failure rates.

case 2: A manufacturer`s declaration is available containing either not the relevant failure rates or no trustworthy ones.

case 3: No manufacturer`s declaration is available.

In order to continue the *PFD* calculation, for all devices meeting cases 2 and 3 reference rates have to be mined. Their significance differs depending on the environmental derivation conditions. Table 1 gives a collection of various reference data.

Table 1 – Reference data

component type	**environment**					
	laboratory		**laboratory-field**		**field**	
	λ_{DU} [FIT]	$\lambda_S+\lambda_{DD}$ [FIT]	λ_{DU} [FIT]	$\lambda_S+\lambda_{DD}$ [FIT]	λ_{DU} [FIT]	$\lambda_S+\lambda_{DD}$ [FIT]
RTD and transmitter	438	2057	700	2100	5670	8505
level sensor and transmitter	18	598	200	550	5731	47279
transmitter power supply	24	143	150	425	2835	0
solenoid driver	0	221	100	400	n/a	n/a
solenoid valve – direct	62	176	585	1510	1400	n/a
solenoid valve – pilot	213	479				
actuator	19	173	670	300	included in valve data	
ball valve	144	143	1350 close 960 open	0	19111 close 14553 open	93439 close 97997 open
					with actuator	
butterfly valve	251	n/a	1600 close 1100 open	0	7317 close 1506 open	37773 close 43584 open
					with actuator	

The given values stem from various data sources [3] – [6]. In some cases adaptations on the environmental conditions in chemical production plants were necessary. In general, three environmental classes are fixed. “Laboratory” contains data derived under laboratory conditions in clean service. Mostly, these failure rates are provided product-specific by manufacturers. “Laboratory-field” consists of data fixed in so-called comprehensive analysis. Based on clean rates the reliability data is modified by experience from onshore field studies. Therefore, these rates represent the boundary between laboratory and field usage. Finally, “field” implies information of big offshore studies which leads to strict field rates. For each class are splitting in rates of dangerous, undetected (λ_{DU}), dangerous, detected (λ_{DD}) and safe failures (λ_S) is annotated. For *PFD* calculations mainly λ_{DU} is relevant because its failure types prevent the system from proper functioning on demand. If the set of component failure rates is completed the *PFD* calculation of the considered loop typical proceeds, aided by tools. The different software solutions can be differentiated by the underlying probabilistic model (Markov Models, formulas of IEC 61508 [7], etc.) and the degree of detail (e.g. additional diagnostic mechanisms). We always used the same tool without publishing its name in order to avoid any commercialism. The underlying proof test intervals are chosen in conformance with the NAMUR data [8] in section 4.

3.2 *PFD* Calculation For Single-channel Loop Typicals

Table 2 shows three single-channel loop typicals which are used by big chemical companies as standard solutions if a pressure-, temperature- or level-monitoring SIF should be realized.

Table 2 – Single-channel typicals

<table>
<tr><th colspan="2"></th><th colspan="3">1oo1 typical</th></tr>
<tr><th colspan="2">process quantity</th><td>pressure</td><td>temp.</td><td>level</td></tr>
<tr><th colspan="2">denotation</th><td>PIRZ+A+</td><td>TIRZ+A+</td><td>LIRZ+A+</td></tr>
<tr><th colspan="2">behavior of final element on demand</th><td>close</td><td>open</td><td>close</td></tr>
<tr><th rowspan="4">sensor part</th><th>sensing element</th><td rowspan="2">X</td><td>O</td><td rowspan="2">X</td></tr>
<tr><th>trans-mitter</th><td>O</td></tr>
<tr><th>power supply</th><td>O</td><td>O</td><td>X</td></tr>
<tr><th>PLC input</th><td>X</td><td>X</td><td>X</td></tr>
<tr><th>logic solver part</th><th>PLC</th><td>X</td><td>X</td><td>X</td></tr>
<tr><th rowspan="5">final element part</th><th>PLC output</th><td>X</td><td>X</td><td>X</td></tr>
<tr><th>solenoid driver</th><td>O</td><td>not used</td><td>not used</td></tr>
<tr><th>solenoid valve</th><td>O
pilot</td><td>O
pilot</td><td>O
direct</td></tr>
<tr><th>actuator</th><td>O</td><td>O</td><td>O</td></tr>
<tr><th>valve</th><td>O
ball</td><td>O
ball</td><td>O
ball</td></tr>
</table>

X Component with sufficient data (case 1 in section 3.1).
O Component with no or insufficient data (cases 2 and 3 in section 3.1).

Their denotations give information about the trip type (high "+" or low "-"). Moreover, the behavior of the final element on demand is listed. The loop itself is divided into sensor-, logic solver and final element part. The component types are assigned to these general groups. Manufacturer and product names are left out as it is not our intention to blame certain manufacturers but to indicate a general problem. Many components suffer from a lack of failure rates, see Table 2. One reason for this phenomenon is the pragmatism of NAMUR [8] companies to use exclusively proven-in-use components for safety-related loops. Before these products are qualified for safety-critical applications, they have to stand laboratory tests and usage periods in the BPCS. Consequently, many proven-in-use components are relatively old. Moreover, failure rates for devices having direct contact with process fluid are problematic in general.

On order to get *PFD*s for the considered loop typicals, missing component rates are replaced by valid reference values, see Table 1. According to different environmental conditions of reference rates, each loop *PFD* is calculated with three different specifications.

PFD_{1oo1}(l) specification	–	missing or insufficient failure data are replaced by reference data derived under laboratory conditions ("laboratory" in Table 1);
PFD_{1oo1}(lf) specification	–	missing or insufficient failure data are replaced by reference data derived under laboratory conditions, modified by knowledge from small onshore field studies ("laboratory-field" in Table 1);
PFD_{1oo1}(f) specification	–	missing or insufficient failure data are replaced by reference data derived from field studies, mostly from offshore applications ("field" in Table 1);

Table 3 mirrors the *PFD* results for the different specifications. These values serve as boundaries of *PFD* bandwidths. The laboratory bandwidth of each typical is given by $[PFD_{MooN}(\mathrm{l}), PFD_{MooN}(\mathrm{lf})]$, the field equivalent by $[PFD_{MooN}(\mathrm{lf}), PFD_{MooN}(\mathrm{f})]$. According to [1] the *PFD* should express the SIF behavior in the field. Hence, only the field bandwidth is valid to serve as realistic for plant application.

Table 3 – *PFD* values of single-channel typicals

1oo1 typical	$\boldsymbol{PFD_{1oo1}}$**(l) [absolute]**	$\boldsymbol{PFD_{1oo1}}$**(lf) [absolute]**	$\boldsymbol{PFD_{1oo1}}$**(f) [absolute]**
total	$2.87 \cdot 10^{-3}$	$1.24 \cdot 10^{-2}$	$9.27 \cdot 10^{-2}$
pressure (P)	$2.01 \cdot 10^{-3}$	$1.20 \cdot 10^{-2}$	$9.48 \cdot 10^{-2}$
temp. (T)	$3.38 \cdot 10^{-3}$	$1.23 \cdot 10^{-2}$	$9.84 \cdot 10^{-2}$
level (L)	$3.21 \cdot 10^{-3}$	$1.28 \cdot 10^{-2}$	$8.48 \cdot 10^{-2}$

3.3 *PFD* Calculation For Dual-channel Loop Typicals

Now, we concentrate on the *PFD* calculation for dual-channel loop typicals. Again three loops are chosen as benchmark. Table 4 shows their functionality and their component connectivity as well as information about the availability of failure rates. Consecutively, replacements by reference values of Table 1 are required in order to continue the *PFD* calculation.

Table 4 – Dual-channel typicals

		1oo2 typical					
process quantity		pressure		temp.		level	
denotation		PIRZ+A+		TIRZ+A+		LIRZ+A+	
behavior of final element on demand		close		close		close	
sensor part	**sensing element**	X	X	O	O	O	O
	trans-mitter			O	O		X
	power supply	X	X				
	PLC input	X	X			X	X
logic solver part	**PLC**	X		X		X	
final element part	**PLC output**	X	X			X	X
	solenoid driver	O	O	O	O	O	O
	solenoid valve	O pilot	O pilot	O direct	O direct	O pilot	O pilot
	actuator	X	O	O	O	O	O
	valve	X globe	O ball	O ball	O ball	O butterfly	O butterfly

X Component with sufficient data (case 1 in section 3.1).
O Component with no or insufficient data (cases 2 and 3 in section 3.1).

Concerning redundant structures not only the quality of failure rates has main influence on the *PFD* outcome. Failure types affecting both channels at the same time must be modeled. They are described by the so-called common cause failure factor β. If β increases, it will dominate the loop *PFD*. IEC 61508 [7] part 6 assists the user in fixing a valid β. For sensor and final element devices values between 1 and 10% are possible. From our point of view a range of 5 to 10% is realistic. To regard β in the following calculations, the *PFD* specifications of section 3.2 are extended.

PFD_{1oo2}(l) specification – like PFD_{1oo1}(l) specification, additionally the common cause failure factor for sensor and final element part is set to 5%;

PFD_{1oo2}(lf) specification	–	like PFD_{1oo1}(lf) specification, additionally the common cause failure factor for sensor and final element part is set to 7.5%;
PFD_{1oo2}(f) specification	–	like PFD_{1oo1}(f) specification, additionally the common cause failure factor for sensor and final element part is set to 10%;

By software-aided *PFD* calculations the results in Table 5 are obtained. Relating to realistic statements, exclusively the field bandwidth [PFD_{MooN}(lf),PFD_{MooN}(f)] claims validity.

Table 5 – *PFD* values of dual-channel typicals

1oo2 typical	**PFD_{1oo2}(l) [absolute]**	**PFD_{1oo2}(lf) [absolute]**	**PFD_{1oo2}(f) [absolute]**
total	$1.62 \cdot 10^{-4}$	$1.10 \cdot 10^{-3}$	$1.34 \cdot 10^{-2}$
pressure (P)	$1.75 \cdot 10^{-4}$	$1.04 \cdot 10^{-3}$	$1.49 \cdot 10^{-2}$
temp. (T)	$1.42 \cdot 10^{-4}$	$1.12 \cdot 10^{-3}$	$1.76 \cdot 10^{-2}$
level (L)	$1.69 \cdot 10^{-4}$	$1.15 \cdot 10^{-3}$	$7.58 \cdot 10^{-3}$

4. FIELD-BASED *PFD* EVALUATION

4.1 General Approach (Top-Down) And NAMUR Data Base

To check if the bottom-up approach is able to map realistic *PFD*s, NAMUR [8] has motivated a data ascertainment within its member companies. In the year 2003 already 33 companies of the chemical and pharmaceutical industry participated, in 2004 and 2005 the group of data suppliers increased to 37. The main advantage of this data base lies in its quality because the material is gained in the field. Consequently, it also includes the observance of the process interface, i.e. the contact with process fluid. Over the years the data structure has become more and more complex. In the initial year 2002 only a distinction between failures in single-channel and multi-channel loops was queried. Since 2003 a subdivision regarding the monitored process quantity has been raising the data complexity. In 2004 a further subdivision into failures in sensor, logic solver and final element part was introduced. Table 6 shows the collected data sets for the three years. Thereby, the following comments should be noticed:
From the statistical point of view a subdivision of the dual-channel loops (1oo2) is not reasonable due to the small number of failures.
Loops with a higher degree of redundancy represent approximately one percentage of the multi-channel loops. Hence, they are not listed.

As not all data suppliers gave a subdivision of their data for the monitored process quantity, the elements of the "total" groups do not meet the sum of the according subgroups. Moreover, groups for "others", "manual" and "quality" do exist but are not considered in this paper. The individuals of each group and the numbers of dangerous, undetected failures are reported. The real proof test intervals have been interpolated from statements of data suppliers.

Table 6 – NAMUR data 2003 - 2005

year	*MooN*	group	loops [absolute]	dangerous, undetected failures [absolute]	T_I [years]
2003	1oo1	total	12,132	41	0.93
		pressure (P)	1,479	11	0.93
		temp. (T)	1,154	1	0.93
		level (L)	1,020	2	0.95
	1oo2	total	10,967	2	0.93
2004	1oo1	total	16,260	43	0.93
		pressure (P)	2,292	18	0.93
		temp. (T)	1,936	5	0.93
		level (L)	1,368	3	0.95
	1oo2	total	10,817	5	0.93
2005	1oo1	total	18,293	56	0.91
		pressure (P)	2,098	17	0.89
		temp. (T)	1,600	5	0.94
		level (L)	1,911	5	0.90
	1oo2	total	9,778	1	0.91

4.2 *PFD* Confidence Interval Estimation

Initially, the following assumptions are made:

1. Only dangerous, undetected failures are *PFD*-relevant, dangerous, detected ones negligible.
2. Failure rates are constant over time.
3. Failure detection and repair is assumed to be perfect during the proof test.
4. $MTTR \ll MTTD_{\mathrm{DU}}$, because normally $MTTR \approx 8\mathrm{h}$ and $MTTD_{\mathrm{DU}} > 0.5\mathrm{y}$.
5. $\lambda_{\mathrm{DU}} T_{\mathrm{I}} \ll 1$.
6. For the NAMUR data it holds: Each safety-related loop can only suffer from at most one dangerous, undetected failure during the observation period of one year.

Deriving *PFD* values from the given data pool requires an adapted *PFD* formula. Under consideration of assumption 1, PFD_{1oo1} can be calculated as the product of relative mean down time by dangerous, undetected failures and the corresponding failure probability.

In order to determine the relative mean down time MDT_{DUrel}, we start at the absolute value MDT_{DU}. Because of assumption 2 failures are equally distributed over the proof test interval T_{I}. Using the statement of perfect proof test coverage (assumption 3), the mean time to detection of dangerous, undetected failures $MTTD_{\mathrm{DU}}$ is half of T_{I}. If additionally the mean time to repair *MTTR* is regarded, MDT_{DU} is set to

$$MDT_{DU} = MTTD_{DU} + MTTR = \frac{T_I}{2} + MTTR \overset{\text{(assumption4)}}{=} \frac{T_I}{2}. \qquad (1)$$

The relative mean down time due to dangerous, undetected failures results by

$$MDT_{DUrel} = \frac{1}{2}. \qquad (2)$$

Under the assumption of an exponential failure distribution the following expression is valid for the required failure probability over T_I:

$$F(T_I) = 1 - e^{-\lambda_{DU} T_I}. \qquad (3)$$

With respect to assumption 5 a linearization is possible:

$$F(T_I) = 1 - \left(1 - \frac{\lambda_{DU} T_I}{1!}\right) = \lambda_{DU} T_I. \qquad (4)$$

Combining (2) and (4) leads to

$$PFD_{1oo1} = \frac{1}{2} \cdot \lambda_{DU} T_I. \qquad (5)$$

Because only failures leading to a complete loss of the systems functionality are reported by NAMUR [8], a dual-channel behaves like a single-channel system.

$$PFD_{1oo2} = \frac{1}{2} \cdot \lambda_{DU} T_I \qquad (6)$$

T_I is directly given in Table 6. λ_{DU} could be estimated be a simple point estimator

$$\lambda_{DU} = \frac{k}{n \cdot \Delta T}, \qquad (7)$$

where ΔT denotes the observation period of one year. But only using (7) for λ_{DU} calculations would not lead to trustworthy values as statistics always imply deficiencies. Estimating confidence intervals compensates for this. Moreover, the resulting *PFD* confidence intervals are easily comparable with the field *PFD* bandwidths of the typicals. The first step to determin λ_{DU} confidence intervals is finding an appropriate interval estimator. Thereby, the choice of the underlying probability distribution plays an important role. As there is no information about the exact time point of failure within ΔT, a continuous model is invalid. Hence, the binomial distribution — as exactest discrete model — delivers good results.

Assuming a binomial distribution, the confidence interval estimator described in [9] can be applied. Initially, the model is transformed from reliability to failure probability estimation. The interval boundaries of the failure probability $p(\Delta T)$ are expressed as

$$p_{low}(\Delta T) = \max_{p(\Delta T)} \sum_{x=k}^{n} \binom{n}{x} (1-p)^{n-x} p^{x} \leq \frac{\alpha}{2} \text{ for } 0 < k < n, \tag{8}$$

$$p_{up}(\Delta T) = \min_{p(\Delta T)} \sum_{x=0}^{k} \binom{n}{x} (1-p)^{n-x} p^{x} \leq \frac{\alpha}{2} \text{ for } 0 < k < n. \tag{9}$$

Thereby, 1- α denotes the confidence level which is to be chosen as at least 70% according to [1]. Solutions of (8) and (9) are determined by iterative methods. We finally get λ_{DU} confidence intervals by usage of the transformation

$$\lambda_{DU} = \frac{p(\Delta T)}{\Delta T}. \tag{10}$$

Table 7 shows the NAMUR [8] *PFD* confidence intervals which have been determined by the illustrated pragmatism.

Table 7 – NAMUR *PFD* confidence intervals

year	*MooN*	group	***PFD*$_{MooN}$ confidence interval** lower boundary [absolute]	upper boundary [absolute]
2003	1oo1	total	$1.32 \cdot 10^{-3}$	$1.87 \cdot 10^{-3}$
		pressure (P)	$2.40 \cdot 10^{-3}$	$4.89 \cdot 10^{-3}$
		temp. (T)	$6.55 \cdot 10^{-5}$	$1.36 \cdot 10^{-3}$
		level (L)	$3.18 \cdot 10^{-4}$	$2.20 \cdot 10^{-3}$
	1oo2	total	$2.90 \cdot 10^{-5}$	$2.00 \cdot 10^{-4}$
2004	1oo1	total	$1.04 \cdot 10^{-3}$	$1.46 \cdot 10^{-3}$
		pressure (P)	$2.78 \cdot 10^{-3}$	$4.77 \cdot 10^{-3}$
		temp. (T)	$6.69 \cdot 10^{-4}$	$2.04 \cdot 10^{-3}$
		level (L)	$4.62 \cdot 10^{-4}$	$2.09 \cdot 10^{-3}$
	1oo2	total	$1.20 \cdot 10^{-4}$	$3.65 \cdot 10^{-4}$
2005	1oo1	total	$1.20 \cdot 10^{-3}$	$1.61 \cdot 10^{-3}$
		pressure (P)	$2.72 \cdot 10^{-3}$	$4.74 \cdot 10^{-3}$
		temp. (T)	$8.18 \cdot 10^{-4}$	$2.49 \cdot 10^{-3}$
		level (L)	$6.56 \cdot 10^{-4}$	$2.00 \cdot 10^{-3}$
	1oo2	total	$7.56 \cdot 10^{-6}$	$1.57 \cdot 10^{-4}$
2003 to 2005	1oo1	total	$1.26 \cdot 10^{-3}$	$1.51 \cdot 10^{-3}$
		pressure (P)	$3.06 \cdot 10^{-3}$	$4.24 \cdot 10^{-3}$
		temp. (T)	$7.58 \cdot 10^{-4}$	$1.54 \cdot 10^{-3}$
		level (L)	$7.36 \cdot 10^{-4}$	$1.56 \cdot 10^{-3}$
	1oo2	total	$7.51 \cdot 10^{-5}$	$1.76 \cdot 10^{-4}$

For each group and year of the NAMUR [8] data a confidence interval is given. Finally, the values for a summarized data set (2003 to 2005) are presented in order to get a higher statistical accuracy and closer interval boundaries. Required precondition of this modus operandi is a sufficient consistency for the data sets of the various years. This property is fulfilled for the single-channel NAMUR [8] data as each subset (pressure, temperature and level) occupies almost the same specific *PFD* interval over the different years, see Fig. 2.

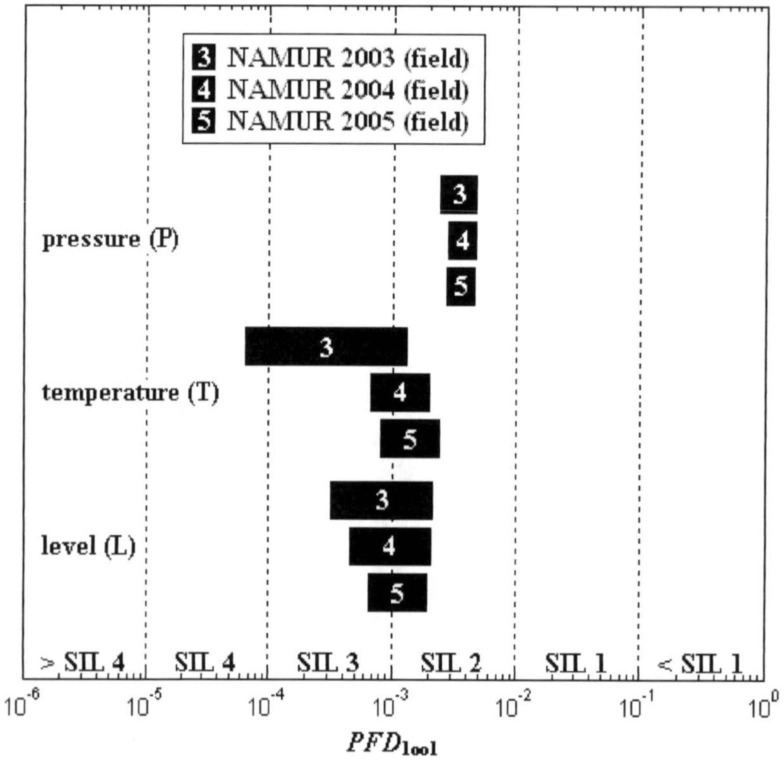

Fig. 2 – Consistency of the single-channel NAMUR data

A similar examination for the dual-channel case does not make sense. About 10,000 loops in combination with small failure numbers suffer from the influence of stochastic effects.

5. CONCLUSIONS

Comparing the *PFD* results of bottom-up (typicals) and top-down approach (NAMUR data) big differences occur. To illustrate the deviations, laboratory and field *PFD* bandwidths are plotted against the field *PFD* confidence intervals. Figure 3 considers the outcome for single-channel loops.

To present the SIL *PFD* ranges properly a logarithmic scaling is chosen. The *PFD* confidence intervals have been generated from a summarized NAMUR data set 2003 to 2005. The *PFD* bandwidth of the "total" typical corresponds to the average bandwidth of the subgroup typicals.

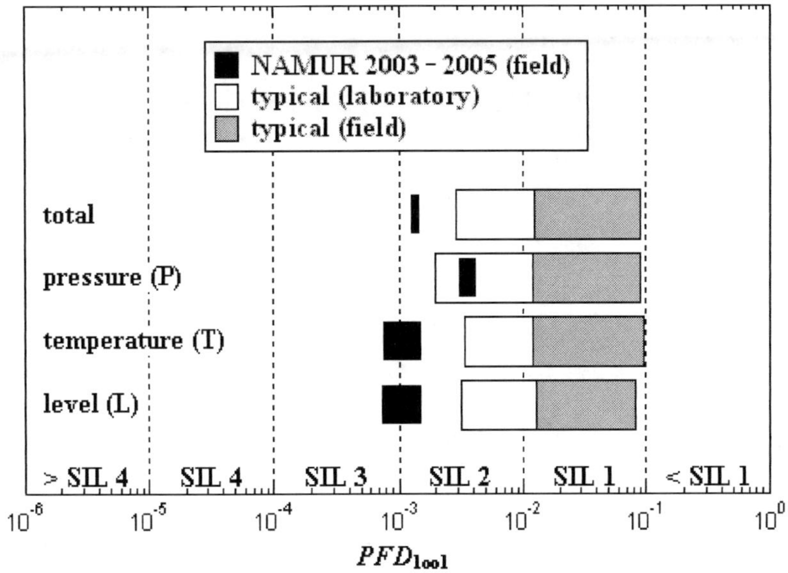

Fig. 3 – *PFD* results of the single-channel loops

Analyzing Figure 3 obeys two important observations:

1st observation: A worst case consideration for the field typical bandwidths (grey) confirms SIL 1 *PFD*s. The field NAMUR confidence intervals (black) indicate SIL 2 quality. This deviation is caused by too conservative input data (i.e. failure rates) for the bottom-up approach. Although there are intersections between NAMUR and typicals, method conformity cannot be derived because field (NAMUR, black) would be mixed up with laboratory data (typicals, white).

2nd observation: The *PFD* typical bandwidths have almost the same location independent from the monitored process quantity and a distinction between laboratory and field values. However, there is a ranking among the NAMUR [8] field data. "Temperature" and "level" show almost congruency, "pressure" is worse. This observation indicates a conflict with the classical *PFD* distribution in IEC 61508 [7] (35% sensor, 15% logic solver and 50% final element part). The final element structures of the safety-related loop typicals always look very similar independent from the monitored process quantity (see Tables 2 and 4). Hence, the ranking among the NAMUR data must originate from the sensor part as main causer of the loop *PFD*. Deeper analyses in [10] confirm this suspicion.

In the two-channel case the NAMUR data do not allow subdivisions. Therefore, an average *PFD* bandwidth of all dual-channel typicals is compared with the *PFD* confidence interval of all dual-channel NAMUR data, see Fig. 4.

An examination of the same criteria as in the single-channel case leads to the following result:

1st observation: The *PFD* deviation between NAMUR data and typical stays almost identical as in the single-channel case. However, all *PFD*s shift one SIL higher.
2nd observation: Analyzing a certain ranking is not possible because there is only one main group.

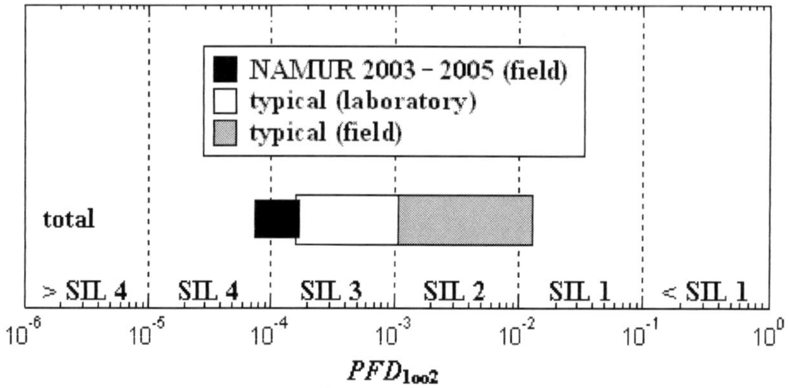

Fig. 4 – *PFD* results of the dual-channel loops

Summing up, the bottom-up approach is not able to map realistic field *PFD* values (at least for Western Europe). To solve this problem, work is in progress to make a further subdivision of the NAMUR data. The aim is to isolate characteristic average λ_{DU}s for sensor, logic solver and final element part. These rates could serve as reference values for proven-in-use components without reliable failure rates. Another embarked strategy is a splitting of manufacturer data according to different environmental temperatures and humidity levels. Most failure rates given by manufacturers are based on worst case assumptions. De facto, these extreme values are not met in chemical production plants. Finally, the rather conservative bottom-up approach should converge against the top-down reality.

6. REFERENCES

[1] IEC 61511, *Functional safety: Safety Instrumented Systems for process industry sector*, Geneva (Switzerland): International Electrotechnical Commission, 2003.

[2] Litz, L., *Safety and Availability of Components and Systems*, in IEEE PCIC Europe Conference Record, 2004, pp 16-21.

[3] Exida.com L.L.C., *Safety Equipment Reliability Handbook*, second edition, Sellerville (USA), 2005.

[4] SINTEF Industrial Management, *OREDA – Offshore Reliability Data*, Hovik (Norway): Det Norsk Veritas, 2002.

[5] Hauge, S., Hokstad, P., *Reliability Data for Safety Instrumented Systems – PDS Data Handbook*, Trondheim (Norway): SINTEF, 2004.

[6] ICI database GEG 3.2.

[7] IEC 61508, *Functional safety of electrical/ electronic/ programmable safety-related systems*, Geneva (Switzerland): International Electrotechnical Commission, 2000.

[8] NAMUR, "Interessengemeinschaft Automatisierungstechnik der Prozessindustrie", http://www.namur.de.

[9] Kumamoto, H., Henley, E. J., *Probabilistic Risk Assessment and Management for Engineers and Scientists*, second edition, New York: IEEE Press, 1996.

[10] Düpont, D., Litz, L., Netter, P., *Sources of Mistakes in PFD Calculations for Safety-related Loop Typicals*, in IEEE PCIC Europe Conference Record, 2007.

Legal Consideration of Changing Industry Standards

There are many OSHA standards which expressly incorporate industry guidance documents and many other OSHA standards which set forth vague and subjective requirements. The operative terms in these standards are words such as "adequate" or "appropriate." OSHA uses industry guidance documents to explain and then enforce these terms. This presentation will discuss the legal issues associated with OSHA enforcement of these terms and associated with adhering to industry standards as they evolve.

Mark S. Dreux
Arent Fox PLLC
1050 Connecticut Avenue, N.W.
Washington, D.C. 20036
dreux.mark@arentfox.com
Phone: (202) 857-6405
Fax: (202) 857-6395

The ATEX Directives: Explosion Safety and Regulation – An Update on the European Approach

N.H.A. Versloot, A.J.J. Klein, M. De Maaijer
TNO Defence, Security and Safety
Rijswijk, The Netherlands
Email : nico.versloot@tno.nl, arjen.klein@tno.nl, matthijs.demaaijer@tno.nl

ABSTRACT

On July 1, 2003 a transitional period has ended and two European directives became fully active:

- *Directive 1999/92/EC*
- *Directive 94/9/EC*

These directives have an impact on both companies with an explosion hazard (gas, vapor, mist or dust explosions) and on manufacturers of equipment intended to be used in explosive atmospheres. The directives had to be incorporated into the national law of each member state of the European Union no later than July 1, 2003. They are examples of achieving a common European market. In this paper, the background of the European ATEX Directives will be discussed and the impact on the users and manufacturers of equipment intended to be used in potentially explosive atmospheres will be highlighted. By adding some experiences after the introduction of the ATEX regulations, this paper forms an update on the paper presented at the 19th Annual CCPS International Conference in Orlando, FL, in 2004.

1. INTRODUCTION

The origin of the European directives dates back to the year 1957 when the European Union was founded. The goal is to achieve a common market for goods, services, people and capital. In fact, trade between member states should be the same as trade within one member state (it is why we now pay with the 'euro' within Europe). This calls for common legislation, because otherwise differences in local legislation could cause differences in competitiveness for a company in selling its products in the various member states. With regard to process safety each member state had its own safety rules to achieve its own safety demand. Although each member state may have had a high demanded safety level the differences were still considerable. Especially in the way this safety level should be reached by the industry. For manufacturers of safety devices and devices to be used in explosive atmospheres this meant that for different member states they had to run different tests and certification procedures in order to be able to sell their products and have them used by their customers. Therefore, harmonization in the safety legislation was needed, according to what had been done in other areas of the common market (e.g. on food quality and safety).

In order to achieve harmonized legislation within the EU, it is normal practice that the European Commission submits a proposal for a new directive to the Council of Ministers. Since the European Community can not prescribe legislation for a member state, it will prepare a directive in which the aspects of the new legislation will be described. It has been agreed between the European member states to implement these directives into their own laws. Effectively, this way the European directives become legislation.
Until 1985, the proposal for new directives had to contain all the technical details, i.e. all test methods and certification procedures intended to be used to put the directive into effect. This provided a major hassle, because this asked for a technical discussion within the non-technical Council of Ministers. Next to that, the only way to have a directive come into force was by a unanimous vote from the Council. This process gave rise to many problems, especially of course on the contents of and agreements on new directives. From 1985 this problem has been solved by changing the nature of the directives: they now give a frame in which the intention of the directive is given and in which references are made to test methods and certification procedures, if necessary. These test methods and certification procedures are then developed by subcommittees. This new process of deriving the European directives is called the "New Approach".

In the field of process safety regarding explosion hazards by gases, vapors and dusts, the EC has come up with two directives:

- Directive 1999/92/EC
- Directive 94/9/EC

These directives are commonly called the ATEX directives (ATEX stands for *ATmosphères EXplosives*) and are normally referred to by their respective article number within the European treaty. So, directive 1999/92/EC is called ATEX 137, directive 94/9/EC is called ATEX 95.
The test methods and certification procedures necessary for the directives are developed by the EC Technical Committee TC305.
Both directives are official legislation, since each member state had to incorporate these directives into law no later than July 1, 2003. For the 1999/92/EC Directive a transitional period existed: employers would have to comply with the Directive no later than July 1, 2006 for existing processes. New processes (i.e. started after June 30, 2003) would have to comply with the directive immediately.

It is important to say that for understanding the requirements following from the Directives, it is always best to check how the Directives have been incorporated into law by the different member states. This is important for the Directive 1999/92/EC (ATEX 137), since this directive gives minimum requirements, which means that it is allowed for member states to add more requirements to the rules prescribed by the directive. The Directive 94/9/EC (ATEX 95) deals with procedures for certifying equipment and as such is a truly harmonized regulation, meaning that it is not allowed for member states to change or add anything to the directive's requirements while implementing it into the member state's law.

2. DIRECTIVE 1999/92/EC (ATEX 137)

The ATEX 137 directive gives '*the minimum requirements for improving the safety and health protection of workers potentially at risk from explosive atmospheres*' [1]. In both directives an explosive atmosphere has following aspects:

1. flammable substances (in the form of gases, vapors, mists or dusts)
2. mixed with air
3. under atmospheric conditions (roughly spoken: temperatures in between -20 and 60 °C and pressures in between 0.8 and 1.1 bar)
4. after ignition, the combustion spreads to the entire unburned mixture

If any of these aspects fail, the atmosphere is considered not to be an explosive atmosphere in the sense of the directive (therefore, it would not fall under the ATEX directives).
Normally, this should not be a problem since most other cases are covered by other legislation. This is why the ATEX directives were not intended to replace any existing legislation, but form an addition to the safety rules already in place. E.g., processes running under high pressure deal with the Pressure Equipment Directive (PED [2]). Again, it should be noted that the directive gives *minimum* requirements, leaving the option to each member state to add more requirements while implementing the directive into law. In fact, some European countries (like Germany and France) have done so.

There are some exceptions for which the ATEX 137 directive is not applicable:

- Areas used directly for and during the medical treatment of patients
- Use of appliances burning gaseous fuels (another directive is in place for this)
- Manufacture, handling, use, storage and transport of explosives or chemically unstable substances
- Mineral-extracting industries (other directives are in place)
- Use of means of transport by land, water and air, with the exception of means of transport intended to be used in a potentially explosive atmosphere

As said, the directive provides requirements for the health and safety protection of workers and is, as such, important for employers. In short, their obligations following from this directive are:

- Prevention of and protection against explosions
- Performing risk analyses
- Area classification to identify (potential) explosive atmospheres
- Use of certified equipment and explosion protection systems
- Writing an Explosion Safety Document

Being a directive of the "New Approach", ATEX 137 will not give any details on how to perform risk analyses or area classifications. For this, it refers to national standards and 'best rules of practice'. In the Netherlands, use is often made of the "Nederlandse Praktijk Richtlijn" NPR7910 (Dutch Practice Guideline), Part 1 for gas explosions, Part 2 for dust explosions [3,4]. There is however a European Guideline, but this is explicitly being given the status 'non-binding' [5]. Still, it is very useful to start with.

The employer needs to investigate the explosion risks by taking into account:

- Normal operating conditions, including maintenance
- Commissioning and decommissioning (like plant startup and shutdown)
- Malfunctions, foreseeable fault conditions
- Misuse which may reasonably be foreseen

with regard to:

- The work equipment used
- The building fabric and configuration
- The substances used
- Work and process conditions and work procedures
- Possible interactions between processes as well as between processes and the working environment
- Places that are or can be connected via openings to places in which explosive atmospheres (may) occur

This can be done by all sorts of methods, varying from systematic questionnaires to FMEA, HAZOP and ETA.
In order to have a situation in which an explosion can have dangerous consequences all following conditions must be fulfilled:

- High degree of dispersion of flammable substance(s)
- Concentration of flammable substance(s) within explosion limits
- Hazardous quantities of explosive atmospheres
- Effective ignition source

So, valid questions would be: do we have any inflammable substances? Can an explosive atmosphere occur? Where will this explosive atmosphere exist? Can we get a *hazardous* explosive atmosphere? Did we do enough to prevent a hazardous explosive atmosphere from occurring? If there will be an explosive atmosphere: did we do enough to prevent ignition of this explosive atmosphere? If ignition of an explosive atmosphere is still possible: did we do enough to prevent dangerous situations resulting from an explosion?
Therefore, an employer should:

- Prevent the formation of an explosive atmosphere
- Prevent the ignition of a hazardous explosive atmosphere (if this atmosphere will occur)
- Reduce the consequences of an explosion to an accepted (non-hazardous) situation

2.1 Explosion Prevention

It should be obvious that prevention of formation of an explosive atmosphere is the first step to take! Countermeasures in this step could be:

- Substituting flammable substances by non-flammable ones

- Reducing the concentration of flammable substances (outside explosion limits)
- Adding inerts to the substances (easier for gases than for dusts)
- Preventing the formation of explosive atmospheres outside process installations by using enclosed installations, ventilation and by preventing the formation of dust layers

The prevention of ignition of explosive atmospheres is based on identification and removal of ignition sources or by assuring that the equipment inside explosive atmospheres can not be an ignition source for that atmosphere. Therefore, the exact places and sizes of (potential) explosive atmospheres should be known. For this, the employer has to carry out an area classification. If an explosive atmosphere may occur locally in such quantities as to require special protective measures to maintain the safety and health of the workers concerned, it is described as a *hazardous* explosive atmosphere. This depends on its volume and the harmful consequences of any ignition. Exceptions to this rule may apply to work with very small quantities, e.g. in laboratories. Therefore, in the Dutch NPR7910 Guideline [3, 4] it is suggested to perform an area classification (or zoning of hazardous places) if a minimum amount of flammable substances is present in a process. This depends on the nature of the substance and the ventilation conditions of the place into which the substance could be released forming an explosive atmosphere. Tables 1 and 2 show the minimum amounts of gases and dusts for which an area classification has to be performed according to the Dutch NPR7910.

Table 1: Minimum amounts of gases for area classification

	Minimum amount
Installations in open air	50 kg
Installations inside buildings	5 kg

Table 2: Minimum amounts of dusts for area classification

	Minimum amount
Inside installations	0.1 kg with particle sizes < 0.1 mm, *or*, 1 kg with particle sizes between 0.1 and 0.5 mm
Inside buildings	50 kg with particle sizes < 0.1 mm, *or*, 500 kg with particle sizes between 0.1 and 0.5 mm

The area classification is based on the likelihood of occurrence of hazardous explosive atmospheres. Explosive atmospheres formed by gases, vapors and mists can be divided into zones 0, 1 and 2; explosive atmospheres formed by dusts can be divided into zones 20, 21 and 22. In Table 3 definitions of these zones are given.

Table 3: Zones in area classification for gases, vapors and mists (zones 0, 1 and 2) and dusts (zones 20, 21 and 22)

Zone	Definition
0	A place in which an explosive atmosphere consisting of a mixture with air of flammable substances in the form of gas, vapor or mist is present *continuously*, or for *long periods* or *frequently*
1	A place in which an explosive atmosphere consisting of a mixture with air of flammable substances in the form of gas, vapor or mist is *likely* to occur in normal operation *occasionally*
2	A place in which an explosive atmosphere consisting of a mixture with air of flammable substances in the form of gas, vapor or mist is *not likely* to occur in normal operation but, if it does occur, will persist for a *short period* only
20	A place in which an explosive atmosphere in the form of a cloud of combustible dust in air is present *continuously*, or for *long periods* or *frequently*
21	A place in which an explosive atmosphere in the form of a cloud of combustible dust in air is *likely* to occur in normal operation *occasionally*
22	A place in which an explosive atmosphere in the form of a cloud of combustible dust in air is *not likely* to occur in normal operation but, if it does occur, will persist for a *short period* only

"Normal operation" means the situation when installations are used within their design parameters.
The Directive does not give a definition for 'continuously', 'long periods', 'occasionally', etc. The Dutch Practice Guideline gives some guidance for this:

Zone 0 or 20: more than 1000 hours/year
Zone 1 or 21: in between 10 and 1000 hours/year
Zone 2 or 22: less than 10 hours/year

Dust layers or deposits will result in either zone 21 or 22 related to their time of occurrence: less than 1000 hours/year for zone 22, more than 1000 hours/year for zone 21 (so, no zone 20 for dust layers).
In the process plant, these zones should be marked by the triangular sign as shown in Figure 1. This sign has black letters wi yellow background (the yellow part should take up at least 50% of

The extent of protective measures depends on the likelihood that hazardous explosive atmospheres will arise (zoning) and should therefore be determined in accordance with Table 4.

Table 4: Extent of protective measures in the various zones

Zones	Ignition sources*) to be reliably avoided
0 or 20	• In normal operation (no malfunction) • In foreseeable cases of malfunction and • In the event of rare malfunctions
1 or 21	• In normal operation (no malfunction) and • In foreseeable cases of malfunction
2 or 22	• In normal operation (no malfunction)

*): in zones 20, 21 and 22 the possibility of deposited dust ignition must also be taken into account

Table 4 applies to all types of ignition sources. Standard EN 1127-1 [6] distinguishes thirteen types of ignition sources:

- Hot surfaces
- Flames and hot gases
- Mechanically generated sparks
- Electrical apparatus
- Stray electrical currents, cathodic corrosion protection
- Static electricity
- Lightning
- Electromagnetic fields in the frequency range from 9 kHz to 300 GHz
- Electromagnetic radiation in the frequency range from 300 GHz to $3 \cdot 10^6$ GHz or wavelength range from 1000 μm to 0.1 μm (optical spectrum)
- Ionizing radiation
- Ultrasonics
- Adiabatic compression, shock waves, gas flows
- Chemical reactions

2.2 Explosion Protection (Mitigation Measures)

If the employer has done his utmost to prevent explosive atmospheres from occurring and igniting and there is still a residual risk on explosions, he should take mitigation or protective measures. Such measures are:

- Explosion-resistant design
- Explosion relief
- Explosion suppression
- Prevention of flame and explosion propagation

Especially the prevention of explosion propagation (explosion decoupling) is strongly highlighted by the Directive.

Furthermore, the employer should focus on organizational measures for explosion protection, like:

- Produce written operating instructions
- Instruct workers in explosion protection
- Ensure workers have adequate competence
- Apply a permit-to-work system for dangerous work
- Carry out maintenance
- Carry out inspection and surveillance
- Mark hazardous places where necessary (by using the sign as shown in Figure 1)

Where independent persons or teams are working simultaneously and in proximity to one another, they may inadvertently place each other at risk, particularly since those concerned focus primarily on their own task, while often nothing or not enough is known about the commencement, nature or extent of the work being done by others nearby. When work is contracted out, the principal and contractor are required to coordinate their activities in order to avoid placing each other at risk. Therefore, employers should take coordination arrangements and provide protective measures to ensure safe cooperation. This should be done for the planning stage, the execution stage and the completion of the work.

2.3 Explosion Protection Document

As one of his duties under the 1999/92/EC Directive the employer must ensure that an Explosion Protection Document is drawn up and kept up to date.
This document must at least demonstrate:

- That the explosion risks have been identified and assessed
- That adequate measures will be taken to attain the aims of the Directive
- Those places which have been classified into zones
- Those places where the minimum requirements set out in Annex II to the Directive will apply
- That the workplace and work equipment, including warning devices, are designed, operated and maintained with due regard for safety
- That arrangements have been made for the safe use of work equipment

In short, the employer should describe his risk analyses, the way he is preventing explosive atmospheres from occurring and how explosion protection is guaranteed, installed and maintained. The document will also contain the organizational and coordination measures.

3. DIRECTIVE 94/9/EC (ATEX 95)

The ATEX 95 Directive deals with '*equipment and protective systems intended for use in potentially explosive atmospheres*' [7]. All products as defined by the Directive which are placed on the market or put into service should comply with this Directive. Products in the scope of the directive are:

- Equipment
    - If intended to be used in a potentially explosive atmosphere
    - Should have its own potential source of ignition
- Protective systems
    - Intended to be installed and used (at least partially) in a potentially explosive atmosphere
- Components
    - Essential to the safe functioning of equipment and protective systems, but,
    - With no autonomous function
- Safety, controlling or regulating devices
    - Are also covered by the Directive if they are situated outside the potentially explosive atmosphere, but necessarily linked to equipment within a potential explosive atmosphere

Equipment is considered to have an autonomous function if it can be safely used to deliver its intended function without the need to add any further parts. It may be, however, that specific instructions for installation and use are to be followed.

Excluded from the 94/9/EC Directive are:

- Medical devices intended for use in a medical environment
- Equipment and protective systems where the explosion hazard originates exclusively from the presence of explosive substances or unstable chemical substances
- Equipment intended for use in domestic and non-commercial environments where potentially explosive atmospheres may only rarely be created, solely as a result of the accidental leakage of fuel gas
- Personal protective equipment (covered by another directive). There are occasions where personal protective equipment with its own potential sources of ignition is intended for use in potentially explosive atmospheres. This type of personal protective equipment should follow the procedures laid down in the 94/9/EC Directive to provide the necessary level of explosion safety
- Seagoing vessels and mobile offshore units together with equipment on board such vessels or units, because they are already covered by the IMO Convention. However, fixed offshore units together with equipment on board, and units and vessels, which are not considered to be seagoing are in the scope of directive 94/9/EC
- Means of transport intended solely for transporting passengers by air, road, rail or water networks, as well as means of transport in so far as such means are designed for transporting goods by air, by public road or rail networks or by water. Vehicles intended for use in a potentially explosive atmosphere will not be excluded

- Equipment designed and manufactured specifically for use by the armed forces or in the maintenance of law and order. Dual-purpose equipment is not excluded

Placing ATEX 95 products on the market means first making available, against payment or free of charge, of products in the EU market, for the purpose of distribution and/or use in the EU.
Since the concept of placing on the market refers only to the first time products are made available in the EU for the purpose of distribution and/or use in the EU, the ATEX directive only covers *new* ATEX products manufactured within the EU and *new* or *used* ATEX products imported from a non-EU country. The person placing the product on the EU market could be the manufacturer, his authorized representative or, if none of them is established in the EU, the importer or any other responsible person.

Figure 2: CE conformity marking

Important to understand is that anyone who puts into service products which are covered by the Directive that have been manufactured by himself for his own use, is considered to be a manufacturer. Therefore, he must fulfill the requirements of the Directive.

The manufacturer has responsibility for:

- Conducting an analysis to conclude if his product is subject to the Directive and which requirements apply
- Design and construction of the product in accordance with the Essential Health and Safety Requirements (EHSRs) laid down in the Directive
- Following the procedures for the assessment of the conformity of the product with the EHSRs laid down in the directive

Certain operations may be subcontracted by the manufacturer to his authorized representative.

The 94/9/EC Directive is one of the European directives dealing with the conformity assessment procedures related to the CE-marking of products. It is the manufacturer's duty to investigate which Directives are applicable to his product(s). After fulfilling *all* the requirements laid down in *all* these directives the manufacturer will mark the product with the CE sign (see Figure 2). This marking then implies fulfillment of all applicable directives.

In the ATEX 95 Directive products have been divided into Groups:

Group I
Comprises equipment intended for use in the underground parts of mines, and to those parts of surface installations of such mines, likely to become endangered by firedamp and/or combustible dust.

Group II
Comprises equipment intended for use in other places likely to become endangered by explosive atmospheres.

These Groups are subdivided into Categories. The way in which this categorization has been developed expresses one of the main distinctions of Group I and II. For Group I, the categorization depends on whether the product is to be de-energized in the event of an explosive atmosphere occurring. For Group II, it depends where the product is intended to be used in and if a potentially explosive atmosphere is always present, or is likely to occur for a long or a short period of time.
Subdivision is as follows:

Group I

- **Category M1**
  Should remain functional for safety reasons when an explosive atmosphere is present and are characterized by integrated explosion protection measures functioning in such a way that:
  - In the event of failure of one integrated measure, at least a second means of protection provides for a sufficient level of safety, or,
  - In the event of two faults occurring independently of each other, a sufficient level of safety is ensured.
- **Category M2**
  Are intended to be de-energized in the event of an explosive atmosphere
  - If de-energizing is not immediately possible, appropriate safety measures should be incorporated

Group II

- **Category 1**
  Should be capable to remain within operational parameters, stated by the manufacturer, and ensuring a *very high level* of protection. This equipment is characterized by integrated explosion protection measures functioning in such a way that:
  - In the event of a failure of one integrated measure, at least a second independent means of protection provides for a sufficient level of safety, or,
  - In the event of two faults occurring independently of each other a sufficient level of safety is ensured.
- **Category 2**
  Should be capable of remaining within operational parameters, stated by the manufacturer, and based on a *high level* of protection. The explosion protection relating to this Category must function in such a way that:
  - A sufficient level of safety is provided even in the event of equipment operating faults or in dangerous operating conditions which normally have to be taken into account
- **Category 3**
  Should be capable of remaining within operational parameters, stated by the manufacturer, and based on a *normal level* of protection. The design of these products must:
  - Provide a sufficient level of safety during normal operation

Categories of Group II are followed by a letter: '**G**' for explosive atmospheres created by gases, vapors and mists, '**D**' for explosive atmospheres created by dusts (e.g. II 2D is equipment from Group II, Category 2, to be used in explosive atmospheres created by dusts). Extension '**G/D**' means that the equipment is suitable for both gas and dust environments (not for hybrid mixtures!).

As has already been said, it is the manufacturer's duty to check if his product has to comply with the ATEX 95 Directive. If so, he should design his equipment or protective system from the point of view of *integrated explosion safety*: preventing the formation of an explosive atmosphere as well as sources of ignition and, should an explosion nevertheless occur, to halt it immediately and/or to limit its effects.
The quality of the integrated explosion safety level should be checked by a risk assessment comprised by following four steps:

- Hazard identification
- Risk estimation
- Risk evaluation
- Risk reduction option analysis

The 94/9/EC Directive also describes the procedures whereby the manufacturer or his authorized representative established within the EU ensures and declares that the product complies with the directive. Only after the successful fulfillment of these procedures the manufacturer will be authorized to put the CE-mark (see Figure 2) on the equipment or protection system (i.e., if the equipment also fulfills all requirements following from any other applicable directives).
There are different Conformity Assessment Procedures depending on the equipment Group and Category. Figure 3 shows the various options.

In Figure 3 following is meant by:

EC type examination

- A Notified Body ascertains and attests that a specimen representative of the production envisaged meets the relevant applicable provisions of the Directive
- The Notified Body will
  - Examine technical documentation and verify that the type has been manufactured in conformity with the technical documentation
  - Perform appropriate examinations and necessary tests

Production Quality Assurance

- The manufacturer will operate an approved quality system for production, final equipment inspection and testing
- A Notified Body will have to approve this quality system and will check the system on a regular basis

Product Verification

- The manufacturer will take all measures necessary to ensure that the manufacturing process guarantees conformity of the equipment with the type as described in the EC type examination
- A Notified Body will carry out the appropriate examinations and tests in order to check the conformity of the equipment, protective system or device

Conformity to type

- The manufacturer ensures and declares that the equipment in question is in conformity with the type as described in the EC type examination certificate
- The manufacturer will take all measures necessary to ensure that the manufacturing process assures compliance of the manufactured equipment or protective systems with the type as described in the EC type examination certificate

Product Quality Assurance

- The manufacturer will operate an approved quality system for final equipment inspection and testing
- A Notified Body will have to approve this quality system and will check the system on a regular basis

Internal Control of Production

- The manufacturer ensures and declares that the equipment satisfy the requirements of the Directive
- The manufacturer will establish technical documentation which will enable the conformity of the equipment with the relevant requirements of the Directive

Unit Verification

- The manufacturer ensures and declares that the equipment or protective system which has been issued with the certificate conforms to the requirements of the Directive
- A Notified Body will examine the individual equipment or protective system and carry out the appropriate tests

Notified Bodies (NoBo) provide the professional and independent judgments, which consequently enable manufacturers or their authorized representatives to fulfill the procedures in order to presume conformity to Directive 94/9/EC. Their intervention is required:

- For issuing of EC type examination certificates, and for inspection, verification and testing of equipment, protective systems, devices and components before they can be placed on the market and/or put into service
- For the assessment of manufacturer's quality assurance system in the production phase

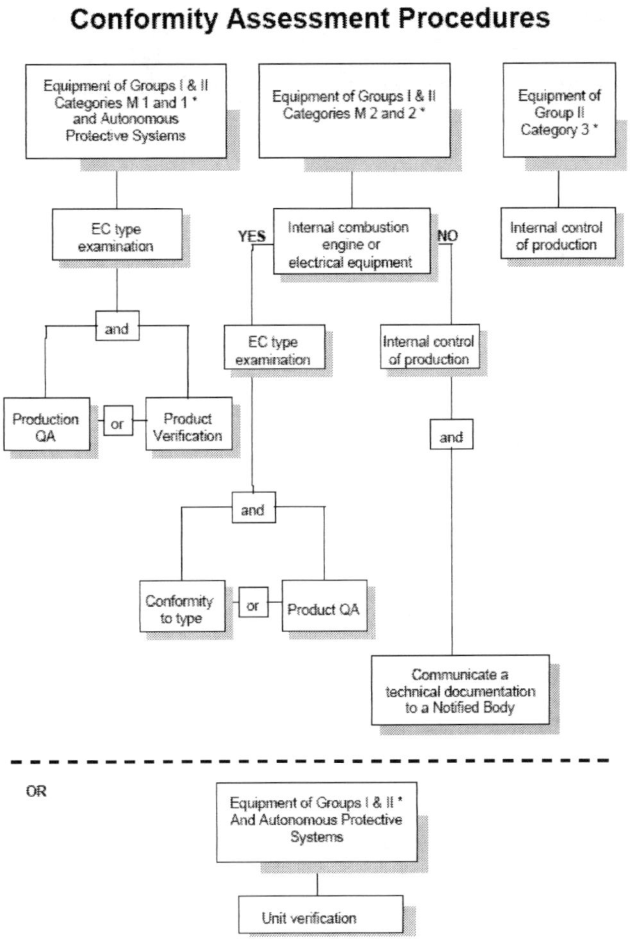

These bodies must be notified by the Member State under whose jurisdiction they fall, on their own responsibility to the Commission and the other Member States of the EU.

So, according to Figure 3 a manufacturer has to involve a Notified Body when dealing with following equipment Group and Category:

- Group I, Categories M1 and M2
- Group II, Category 1 and 2

For Categories M2 and 2 the involvement of the NoBo depends on the type of equipment: if it is an internal combustion engine or electrical equipment or not. In the latter case, the manufacturer only needs to send the technical documentation to the Notified Body.

Figure 3: Conformity Assessment Procedures according to Directive 94/9/EC (Figure taken from page 29 of reference [8])

In order to perform appropriate examinations and tests the Notified Body will make use of standards. The Notified Bodies have internally agreed to consider the following documents when assessing the conformity of a product to the Essential Health and Safety Requirements (EHSRs). The order of preference is from 1 down to 10 (with number 1 being the highest ranking!):

1. Applicable harmonized standards
2. CENELEC/CEN published 2nd edition
3. Applicable EN standards 1st edition
4. Final EN projects ex: EN50014 pr A1
5. EN projects
6. Applicable National standards
7. Interface documents CLC/CEN (if any)
8. Applicable ISO/IEC documents
9. Local testing procedures from a Notified Body
10. Other available standards or part of standards (USA, Japan, etc.)

A useful guide for applying the 94/9/EC Directive is the "ATEX Guideline" [8].

4. LINK BETWEEN DIRECTIVE 1999/92/EC AND DIRECTIVE 94/9/EC

The main link between the two directives comes from the zoning (hazardous area classification) and the equipment Group and Category classification as described in Annex IIB of the 1999/92/EC Directive. Based on the zone and equipment type definitions Table 5 shows which type of equipment may be used in the different zones.

Table 5: Different equipment types to be used in the various zones

Zones	Equipment type
0	1G
1	1G or 2G
2	1G, 2G or 3G
20	1D
21	1D or 2D
22	1D, 2D or 3D

In fact, Table 5 combines the 'worlds' of the manufacturer and the end user: based on the zoning prepared by the end user, the manufacturer is capable of delivering the right type of equipment or protective system.

5. EXPERIENCES AND EVALUATION

Since the transitional period for the 1999/92/EC Directive (ATEX 137) would end on July 1, 2006 it was investigated in the Netherlands if industry would be able to comply with the directive in time. It appeared that not all industrial sectors knew enough about the regulations. In particular, SMEs (with less than 10 employees) had a considerable lack of knowledge. Lack of knowledge and expertise and the fact that the new regulations were not clear to them, were constraining factors in the implementation of the new directives. Other constraints are the additional costs following from applying to the new regulations. These costs are made for:

- Investments made in new certified equipment
- Research into explosion characteristics of gases and dusts
- Development of new (working) procedures

It appeared that the situation in the Netherlands was not different from other European countries. It was expected that the majority of industrial sectors working with explosive gases and dusts would not be able to meet the deadline of implementation of the directives.

This is of particular importance since the Dutch Labor Inspectorate (as part of the Ministry of Social Affairs and Employment) announced more preventive checks on industry for compliance with the new regulations. With this, the Inspectorate shifted its focus from mainly investigating on incidents after they have occurred.

Therefore, it was concluded that a more extensive publicity program should be developed, e.g. through the media and by courses, workshops and documentation. By this, the Ministry would take up more on their role of organizing and controlling a proper implementation of the directives into the Dutch society.

6. FURTHER INFORMATION

More information on the Directives, harmonized standards and Notified Bodies can be found on the EU website: europa.eu.int/comm/enterprise/atex.

From here the Directives and the Guidance to the Directives can be downloaded.

7. NOMENCLATURE

ATEX	*AT*mosphères *Ex*plosives
EC	European Community
EHSRs	Essential Health and Safety Requirements
ETA	Event Tree Analysis
EU	European Union
FMEA	Failure Mode and Effects Analysis
HAZOP	Hazard and Operability Study
NoBo	Notified Body
PED	Pressure Equipment Directive
SME	Small and Medium Enterprise

8. REFERENCES

1. *Directive 1999/92/EC of the European Parliament and of the Council of 16 December 1999 on minimum requirements for improving the safety and health protection of workers potentially at risk from explosive atmospheres (15th individual Directive within the meaning of Article 16(1) of Directive 89/391/EEC)*, Official Journal of the European Communities, No L23, January 28, 2000
2. *Pressure Equipment Directive 97/23/EC*, into force on November 29, 1999
3. *Nederlandse Praktijk Richtlijn NPR7910-1:2001*, Dutch Practice Guideline, Part 1: Gas explosions, based on NEN-EN-IEC 60079-10
4. *Nederlandse Praktijk Richtlijn NPR7910-2:2001*, Dutch Practice Guideline, Part 2: Dust explosions, based on NEN-EN 50281-3
5. European Commission, DG Employment and Social Affairs; Health, Safety and Hygiene at Work, *Non-binding Guide of Good Practice for implementing of the European Parliament and Council Directive 1999/92/EC on minimum requirements for improving the safety and health protection of workers potentially at risk from explosive atmospheres*, Final Version, April 2003
6. *EN 1127-1:1997*, Explosive Atmospheres – Explosion Prevention and Protection – Part 1: Basic concepts and methodology
7. *Directive 94/9/EC of the European Parliament and the Council of 23 March 1994 on the approximation of the laws of the Member States concerning equipment and protective systems intended for use in potentially explosive atmospheres*, Official Journal of the European Communities, No L100, April 19, 1994
8. *ATEX Guidelines (First Edition), Guidelines on the Application of Council Directive 94/9/EC of 23 March 1994 on the Approximation of the Laws of the Member States concerning Equipment and Protective Systems intended for use in Potentially Explosive Atmospheres*, May 2000

Retrofitting New Standards into Existing Facilities is Not Always as Easy as it Seems

Brian R. Dunbobbin, Shakeel H. Kadri, Timothy J. Mc Lear and Richard O'Reilly
Air Products & Chemicals Inc.
7201 Hamilton Blvd.
Allentown, PA 18051 USA
dunbobbr@apci.com

Process Safety related safety standards are often updated, or new standards are developed, following incidents, near misses, or new advances in the safe design of chemical facilities. Whether these new standards should be implemented into existing facilities then becomes an issue that the Process Safety, Engineering and Operating communities need to address. On the surface it would appear that all new safety standards should be applied to existing facilities. However, sometimes the answer is not intuitively obvious because the existing facility may have a very different design approach and design basis to newer facilities producing similar products; therefore it may not be practical, or safe, to blindly introduce new standards into older, existing facilities.

This paper outlines an approach to ensuring that practical operability issues and process safety risks are adequately accounted for when considering the introduction of new standards. Of particular importance are having the issues reviewed by senior Technical and Operations managers' familiar with the design and operation of both the new and the existing facilities.

New engineering standards should not be automatically retroactively applied to existing facilities. During the writing stages of the standard, both Engineering and Operations personnel are actively involved in the standards writing to ensure that practical and safe operation are assured for both new and existing facilities. The final standard may have one design approach for new facilities and several potential approaches for existing facilities depending upon the design. Once the standard has been approved by the writers, it is then subject to review by a Retrofit Standards Committee that ensures that the correct mix of skills was involved in the writing of the standard and that adequate layers of protection are applied to systems that cannot implement new standards because of design or installation differences. This committee requires that the retrofit decision, and any requirements, be considered and be documented during the writing of the standard, taking into account incidents, near misses, etc.

1.0 Introduction

Chemical manufacturing companies that have multiple facilities inevitably face the problem of whether or not to retrofit new or revised standards into their existing facilities. On the surface it would appear that all new safety standards should be applied to existing facilities. However, sometimes the answer is not intuitively obvious because the existing facility may have a very different design approach and design basis to newer facilities producing similar products; therefore it may not be practical, or safe, to blindly introduce new standards into older, existing facilities. In some cases an alternate design can be developed that satisfies the safety issues and ensures a cost effective, practical solution. In other cases administrative controls can offer adequate alternatives.

The key is to recognize that not all changes to standards that apply to new facilities need to be applied to existing facilities. Some changes will have no safety impact. Some will only require key elements to be retrofitted. In the cases where retrofitting is necessary it is important to recognize the basic intent of the modifications to the standard. It is this basic intent that has to be retrofitted and there may be several different ways to achieve that intent.

To demonstrate the use of an alternate design approach rather than the retrofitting of a new standard, let me give an example. Several years ago a standard was written requiring internal shutoff valves on certain hazardous material storage tanks. It was deemed necessary to retrofit this standard onto existing storage tanks in similar service. Draining the existing tanks of the hazardous fluids, removing existing piping and installing internal shutoff valves was impractical, expensive and exposed the installation technicians to potential hazards while they were performing the work. Instead of blindly installing the internal shutoff valves on these existing storage systems the team stepped back and looked at the safety function that the internal valves were performing. This function was to prevent external forces (say a truck) from crashing into the external valve and/or piping and shearing it off up-stream of the shutoff valve, potentially allowing total loss of containment of the tank through the sheared piping. It was decided to instead utilize the existing external valves, install remote shutoff actuators and add a substantial protective "cage" around the external valve. These measures achieved the goal of ensuring a safe operation that could remotely shutoff the flow from the tank and could prevent external forces from damaging the piping upstream of the shutoff valve. New installations typically now utilize the internal shutoff valves while existing facilities utilize the alternate design. An equal level of safety is accomplished in both cases.

This paper outlines how we approach making the retrofit/no retrofit decision. This approach is utilized in the application of new and revised standards to existing facilities, new projects currently underway and to acquisitions that may be made in the future.

2.0 A Work Process for Identifying and Resolving Retrofit Issues

New engineering standards are not automatically retroactively applied to existing facilities. During the writing stages of a standard, both Engineering and Operations personnel are actively involved in the standard writing to ensure that practical and safe operation are assured for both new and existing facilities. The final standard may have one design approach for new facilities and several potential approaches for existing facilities depending upon the design. Once the standard has been approved by the writers, it is then subject to review by a Retrofit Standards Committee that ensures that the correct mix of skills was involved in the writing of the standard and that adequate layers of protection are applied to systems that cannot implement new standards because of design or installation differences. This committee requires that the retrofit decision, and any requirements, be considered and be documented during the writing of the standard, taking into account incidents, near misses, etc.

Standards Coordinators ensure that standards are identified as needing review by the Retrofit Standards Committee. The Coordinators have compiled and maintain a list of standards that have required retrofitting in the past so that this list is available to Engineering, Operations and teams involved in acquiring facilities. This list is, by necessity, not all-inclusive, it is a starting point and is kept evergreen. Not all standards will have to be retrofitted completely, in some cases alternative ways to mitigate the risk may be advantageous for existing facilities e.g., administrative controls.

Temporary installations and operating and maintained installations are subject to the same review mechanism as owned and operated facilities.

3.0 Implementation of the Work Process

Figure 1 is a flowchart which summarizes some of the key items to be considered in making the decision on whether or not to retrofit new or revised standards, or identified requirements from a standard, into existing or acquired facilities.

Regulatory issues are always retroactively implemented into existing facilities. Changes in codes are evaluated by experts on the code in question to ascertain if the code is one that we follow in the facilities in question, if so then it is further assessed for retrofit requirements. Note that not all codes are followed because codes can vary by country or industry, for instance ASME codes are often followed in the US, but not everywhere in Europe or the rest of the world.

A further screen on retrofitability is whether or not the changes in the standards have EH&S impact. If they don't, then this work process is not followed and the decision is left to the Engineering and Operations implementation teams who make the assessment on retrofitting on a purely cost benefit analysis.

If the change in a standard can lead to the possibility that existing facility designs are deficient and could lead to safety consequences then the safety consequences need to be evaluated to ascertain if additional layers of protection are required. Often this assessment is qualitative, but a level of protection analysis (LOPA) may be required. At this stage alternative solutions may have to be developed to ensure safe operation of the facility.

Once the decision is made to retrofit a standard, or key requirements from a standard, into an existing facility it is also necessary to determine a timeframe over which the facility can operate until the modifications are made. This may also require a decision on whether or not temporary operating guidelines need to be issued, and be followed, until the modifications have been completed.

Retrofit requirements for existing facilities are the most obvious retrofit areas but we should not forget two other areas that deserve attention:

1) Current Engineering Projects:

If a new or revised standard requiring retrofitting is in development during the course of a project, or is issued after project kickoff, project management has to decide if they want to make the change during the project phase, make the change after startup, or implement alternate layers of protection that follow the intent of the change and fulfill the safety requirements. This can be a difficult decision since it may impact project schedules and costs dramatically. If required items are not retrofitted before startup, this then would be treated as a retrofit to an existing facility and if necessary additional safety levels of protection may be applied to the facility until the items have been retrofitted. For instance, administrative controls can be implemented to add the required layers of protection.

2) Acquisitions:

During the Pre-Purchase EH&S Assessment phase of an acquisition an assessment has to be made as to the need to retrofit applicable standards into the facility being acquired. If retrofitting is difficult, then alternate design or operational modifications need to be outlined. This is important since the cost of implementing these modifications onto an existing acquired facility can be substantial and need to be factored into the cost to purchase the facility.

4.0 Conclusion

The determination of whether or not to retrofit new or modified standards into existing facilities is not always simple and requires review by experienced engineering, operations and process safety personnel. Existing facilities may have a very different design approach and design basis to newer facilities producing similar products; therefore it may not be practical, or safe, to blindly introduce new standards into older, existing facilities.

A standing Retrofit Standards Committee that reviews new or modified standards that have EH&S impact has proved very efficient in resolving issues related to retrofit decisions. This committee ensures that the correct experts are involved in the writing and review of such standards. They also ensure that alternative design and operational solutions are considered when attempting to retrofit a standard into existing facilities – it is the safety intent of the standard that needs to be retrofitted, and that can sometimes take a very different implementation approach in existing facilities to that taken in new facilities.

Figure 1

Flowchart summarizing key items to be considered when deciding whether or not to retrofit new or revised EH&S related standards

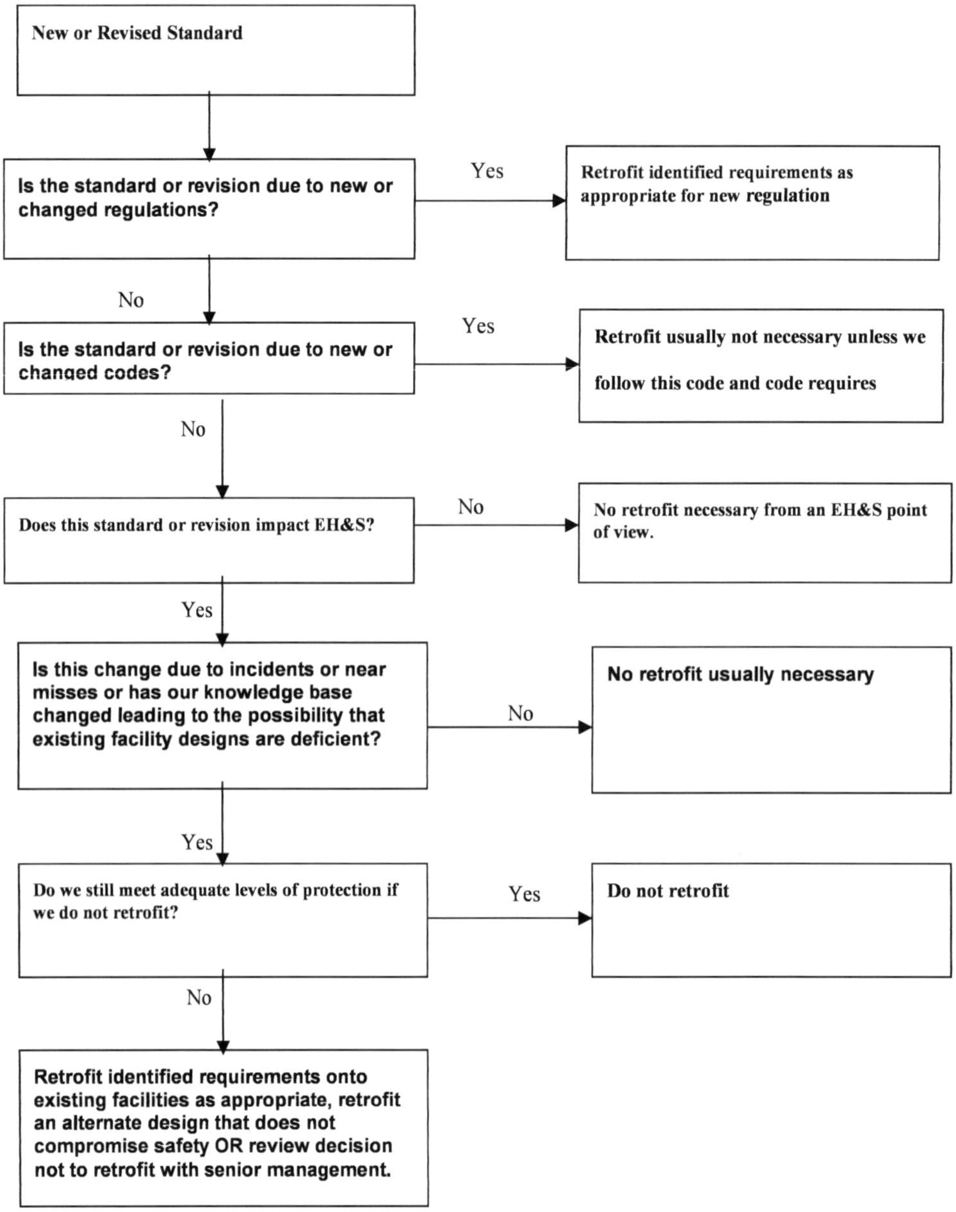

Reactivity Issues in Storage: A Simplified Approach

J. Wayne Chastain, Eastman Chemical Company
William W. Doerr, Ph.D., FM Global
Scott Berger, Center for Chemical Process Safety
Peter N. Lodal, Eastman Chemical Company

Abstract

Chemical reactivity issues, with their potential for generation of heat, overpressure and/or harmful chemical reaction products, continue to present a significant threat for many aspects of chemical commerce (1). While reactivity concerns routinely exist in areas where chemicals are intentionally mixed and processed, these same concerns often occur in simple storage as well. However in storage areas, chemical reactivity issues aren't recognized as often as in other areas.

This paper will discuss simplified reactivity evaluation guidance that is primarily intended for those who store chemicals. Using DOT/UN Shipping Classifications and supplemental material from the NOAA Chemical Reactivity Worksheet and Health & Safety Executive HSG71 "Chemical Warehousing – the storage of packaged dangerous substances", separation recommendations to mitigate or eliminate reactivity issues are suggested.

Although this guidance is targeted for companies whose primary business is chemical storage, this information could also be useful to those businesses where the storage activities also occur as a part of a more extensive manufacturing operation, which could include both physical processing of materials (i.e., distillation, blending, drying, etc.) as well as intentional reactive chemistry operations (manufacturers, toll converters, etc.)

Introduction

The Reactivity Management Roundtable (RMR) was formed as a special project of the Center for Chemical Process Safety (CCPS) in April 2004. The entire scope of the RMR's efforts to promote effective management of reactivity hazards is covered by Berger (2). Other hazards such as flammability, combustibility, toxicity, explosivity, and infectious or radioactive substances are not addressed in this approach, but are addressed in other standards, regulations, or guidelines provided by jurisdictional or industrial organizations such as the EPA, OSHA, NFPA, AIChE CCPS, FM Global, UK H&SE, etc. It is not the purpose of this effort to address these other hazards.

Since the issue of chemical reactivity can be found in the entire chemical manufacturing and distribution process, the RMR group decided to address the issue in key parts of the chemical supply chain: reaction processing, mixing, and storage or warehousing. This effort addresses reactivity issues in venues where chemicals are stored, and where reactivity concerns may be adequately addressed by physical separation and/or isolation with the intent to reduce the potential for ignition from reaction, or escalation of a hazardous situation due to proximity of incompatible materials. The team working on this problem set two guiding principles to direct their work;

1. The method had to be simple and straightforward.
2. The evaluation should be able to be completed with information that the customer already had in hand, without need for additional testing or modeling.

The customers for products of this effort were identified to be:

1. Chemical warehousers and distributors, such as those companies who belong to the National Association of Chemical Distributors (NACD);
2. Chemical producers who maintain chemical inventories either physically or functionally separate from their manufacturing ; or,
3. Public warehouses who accept chemicals for storage
4. Chemical manufacturers who store and warehouse chemicals that they use or manufacture

Thus, the customer base would have familiarity with chemicals and their fundamental properties, but not necessarily chemical reactivity testing, modeling or calorimetry expertise.

Approach

The approach taken was adapted from a method recommended by the UK Health and Safety Executive document "Chemical Warehousing – the storage of packaged dangerous substances" (3). This method uses DOT/UN shipping classifications (both primary and secondary) as a means of determining the nature of the hazard(s) of a chemical and where potentially harmful interactions or self-reactivity hazards may exist, and controls the reactive hazard by separation and/or isolation of the chemical from other chemicals. Five separation categories are used in the RMR approach:

Category	Description of Separation or handling
1.	No special separation is needed
2.	Separate by one aisle or 10 ft. (3 m), whichever is greater
3.	Separate in different compartments with minimum 30 minute firewall
4.	Isolate materials in separated (detached) storage areas.
5.	Extremely Hazardous Material. Special storage/segregation requirements. See applicable references (8)

Advantages and Disadvantages

This approach has two strong advantages:

1 It uses information that is universally available for chemicals in commerce, i.e., the DOT/UN shipping classification. The Material Safety Data Sheet (MSDS) or International Safety Card (ISC) for these materials will contain the shipping classification.

2 It provides easy-to-understand visual information on how materials should be separated; making the storage of materials consistent with, and as safe as or safer than the material as it is shipped.

Disadvantages include:

1. The method does not evaluate chemical interactions explicitly, as more quantitative methods would provide.
2. Known deficiencies in UN/DOT Classification are not corrected within the methodology. One well-known example of such a deficiency is chlorine, which does not have a secondary classification as an oxidizer (5.1) but which is well known as an oxidizer.
3. Quantity of material, either collectively or in individual shipping units, is not a factor in the analysis. The source term in some cases of an accidental release may increase the risk.

The Matrix

The evaluation matrix was taken from the HSE Warehousing Guide (2) and augmented with information from other sources, primarily NFPA guidance documents (4-7). Unlike the HSE guide, the matrix includes all primary categories of shipping classifications. The evaluation matrix uses binary combinations of chemicals based upon their shipping classifications. Rows and columns represent the two chemicals, and the cell in the matrix represents the separation category as noted above. The matrix is shown in Figure 1, which may be viewed in full color in the enclosed CD.

Figure 1[abc]

DOT/UN Category → ↓	1.0	2.1	2.2	2.3	3	4.1	4.2	4.3	5.1	5.2	6.0	6.1	6.2	7.0	8.0	9.0	Non-Haz	Unknown
1.0 Explosives	4	4	4	4	4	4	4	4	4	4	4	4	4	4	4	4	0	3
2.1 Flammable gas	4	0	1	1	2	2	2	2	3	3	1	1	4	4	1	1	0	3
2.2 Non-flammable gas	4	1	0	1	1	0	2	0	0	2	0	0	4	4	1	1	0	3
2.3 Toxic gas	4	1	1	0	2	1	2	1	0	2	0	0	4	4	1	1	0	3
	4	2	1	2	0	1	2	2	3	3	1	1	4	4	1	1	0	3
	4	2	0	1	1	0	1	2	3	2	1	1	4	4	1	1	0	3
4.2 Spontaneously Combustible	4	2	2	2	2	1	0	1	3	3	1	1	4	4	1	1	0	3
	4	2	0	1	2	2	1	0	3	2	0	0	4	4	1	1	0	3
	4	3	0	0	3	3	3	3	0	3	1	1	4	4	1	1	0	3
5.2 Organic Peroxides *(SEE NOTE 3)*	4	3	2	2	3	2	3	2	2	0	1	1	4	4	1	1	0	3
	4	1	0	0	1	1	1	0	1	1	0	0	4	4	1	1	0	3
	4	1	0	0	1	1	1	0	1	1	0	0	4	4	1	1	0	3
	4	4	4	4	4	4	4	4	4	4	4	4	4	4	4	4	0	3
7.0 Radioactive Materials	4	4	4	4	4	4	4	4	4	4	4	4	4	4	4	4	0	3
8.0 Corrosive Substances	4	1	1	1	1	1	1	1	1	1	1	1	4	4	1	1	0	3
	4	1	1	1	1	1	1	1	1	1	1	1	4	4	1	1	0	3
	4	0	0	0	0	0	0	0	0	0	0	0	0	0	0	0	0	3
Unknown Classification	3	3	3	3	3	3	3	3	3	3	3	3	3	3	3	3	3	3

KEY	CODE
No special separation needed	0
Separate by one aisle way or 10 ft. (3 m), whichever is greater	1
Separate in different compartments with minimum 30 minute firewall in between	2
	3
Extremely Hazardous Material. Special storage/segregation requirements. See applicable references	4

[a] NFPA 430 and 432 as well as FM Global Datasheets 7-82 and 7-80 provide guidance on separation or detached storage requirements for certain oxidizers or organic peroxides respectively by chemical type, quantity, type package unit, or in the presence of other chemicals.

[b] Except for identical materials along the matrix diagonal, combinations of similar class materials along the matrix diagonal with code "0" may be incompatible, and hence Code levels greater than Code "0" may be desirable.

[c] Category 9 materials can have some unique hazards. Code levels higher than Code "1" may be desirable in some cases.

Features of note in the matrix are:

1. For shipping categories 8 and 9, a minimum separation of category 2 (10 feet separation) is required except for identical materials. This is conservative, as many compounds in this category are wholly compatible with one another. However, since Category 8 (corrosives) includes both strong acids and strong bases, a minimal level of separation is necessary to avoid any potential incompatibilities.
2. For extremely hazardous materials, as defined by shipping categories 1.0 (explosives), 6.2 (infectious substances) and 7.0 (radioactive substances), the user is referred to other, more specialized guidance. The HSE document does not include these explicitly in their matrix, but gives the same guidance towards external, specialized information. The assumption here is that those who handle these materials will be well aware of the specialized nature of their business, and know that detailed guidance or regulatory/legal requirements exist for these materials.
3. Categories with self-reactive materials (5.1 oxidizers, and 5.2, peroxide forming materials) trigger a special warning note to delve deeper into the specialized information contained in the MSDS or other specialized guidance (see footnotes with Figure 1).

The matrix is used to compare the four combinations of shipping classifications for binary pairs of chemicals (primary-primary, primary-secondary, secondary-primary, and secondary-secondary) and choose the highest separation category indicated for these interactions as the suggested storage configuration. Putting the matrix into a spreadsheet or database format allows for automated comparison of all binary pairs of a given chemical list.

An Example

Figure 2 shows a hypothetical example of a chemical warehousing application for a multi-purpose chemical manufacturing facility. The information in this figure is taken directly from the MSDS of the materials listed, all of which are readily available.

Figure 2

Chemical Listing for Example Warehousing Application

	CAS number	DOT UN Classification (Primary)	DOT UN Classification (Secondary)
Sulfuric Acid, 93%	7664-93-9	8.0	
Sodium Hydroxide, 50% solution	1310-73-2	8.0	
tert-Butyl Hydroperoxide, 70%	75-91-2	5.2	8.0
Sodium Hydrosulfide, 18-22%	16721-80-5	8.0	
Isopropyl Chloroformate	108-23-6	3.0	
Dimethyl Sulfate	77-78-1	6.1	
Phosphorus Trichloride, 90 - 100%	7719-12-2	6.1	8.0
Ethylene Oxide	75-21-8	2.3	2.1
Hydrochloric Acid, 33 - 40%	7647-90-5	8.0	
Water	7732-18-5	Non-Haz.	
Triethyl Aluminum		4.2	4.3
Aluminum, powder	7429-90-5	4.3	
Ethanol, denatured, 200 proof	64-17-5	3.0	
Chlorine	7782-50-5	2.3	
Sodium Hypochlorite, 5-15% solution	7681-52-9	8.0	
Aqueous Ammonia, 10% solution	1336-21-6	8.0	

Figure 3 shows the interaction matrix that result from these 16 compounds.

	Figure 3 Example Chemical	CAS number	DOT UN Classification (Primary)	DOT UN Classification (Secondary)	1	2	3	4	5	6	7	8	9	10	11	12	13	14	15	16
1	Sulfuric Acid, 93%	7664-93-9	8.0		0	1	1	1	1	1	1	1	1	0	1	1	1	1	1	1
2	Sodium Hydroxide, 50% solution	1310-73-2	8.0		1	0	1	1	1	1	1	1	1	0	1	1	1	1	1	1
3	tert-Butyl Hydroperoxide, 70%	75-91-2	5.2	8.0	1	1	0	1	3	1	1	3	1	0	3	2	3	2	1	1
4	Sodium Hydrosulfide, 18-22%	16721-80-5	8.0		1	1	1	0	1	1	1	1	1	0	1	1	1	1	1	1
5	Isopropyl Chloroformate	108-23-6	3.0		1	1	3	1	0	1	1	2	1	0	2	2	0	2	1	1
6	Dimethyl Sulfate	77-78-1	6.1		1	1	1	1	1	0	1	1	1	0	1	0	1	0	1	1
7	Phosphorus Trichloride, 90 - 100%	7719-12-2	6.1	8.0	1	1	1	1	1	1	0	1	1	0	1	1	1	1	1	1
8	Ethylene Oxide	75-21-8	2.3	2.1	1	1	2	1	2	1	1	0	1	0	2	2	2	1	1	1
9	Hydrochloric Acid, 33 - 40%	7647-90-5	8.0		1	1	1	1	1	1	1	1	0	0	1	1	1	1	1	1
10	Water	7732-18-5	Non-Haz.		0	0	0	0	0	0	0	0	0	0	0	0	0	0	0	0
11	Triethyl Aluminum		4.2	4.3	1	1	3	1	2	1	1	2	1	0	0	1	2	2	1	1
12	Aluminum, powder	7429-90-5	4.3		1	1	2	1	2	0	1	2	1	0	1	0	2	1	1	1
13	Ethanol, denatured, 200 proof	64-17-5	3.0		1	1	3	1	0	1	1	2	1	0	2	2	0	2	1	1
14	Chlorine	7782-50-5	2.3		1	1	2	1	2	0	1	1	1	0	2	1	2	0	1	1
15	Sodium Hypochlorite, 5-15% solution	7681-52-9	8.0		1	1	1	1	1	1	1	1	1	0	1	1	1	1	0	1
16	Aqueous Ammonia, 10% solution	1336-21-6	8.0		1	1	1	1	1	1	1	1	1	0	1	1	1	1	1	0

Some things to note about this example:

1. The matrix allows storage of water with all compounds. This is a result of its non-hazardous classification. There are at least two compounds that are violently reactive with water and these are identified as shipping classification 4.3 (dangerous when wet); thus, because the matrix does not handle water properly, the user **must** examine further any shipping classification "4.3", against classification 8.0 for water as well as any chemical descriptor with "aqueous", "solution", or "%" appropriate when water solution in the term because water reactivity should be assessed for each individual compound. These problematic binaries are noted with BLUE background (when 4.3 with water as 8.0) and RED category code (when water is implicit by the descriptor.) In these cases (BLUE background or RED category code), the separation category should be at least "1".
2. For other compounds, which are categorized corrosive, there is a minimum of 10 feet separation recommended. For the case of compounds 15 and 16 (aqueous sodium hypochlorite and aqueous ammonia) this is clearly conservative.
3. Chlorine, a known oxidizer, does not carry this designation as either a primary or secondary shipping classification. This is a deficiency in its classification, and could be handled by adding a "pseudo"-shipping classification. Similar issues could occur with other chemicals, or the same chemical with different MSDSs.

Future Enhancements

While this simplified tool provides a simple, useful look at reactivity in storage applications, there are several improvements that are planned to increase its utility. These include:

1. Use of an automated MSDS retrieval system to facilitate data entry. Linking to either university (e.g. University of Vermont) or commercial (e.g. MSDSOnline) sites that provide MSDSs would not only improve data input accuracy and decrease input time, it would automatically link the MSDS to the tool. For those cases where more detailed information is needed, the MSDS would provide a ready reference to either the pertinent data, or a source where additional information could be obtained.
2. Development of a more detailed protocol to evaluate non-storage situations (i.e. physical mixing and intentional chemistry applications. This is a natural extension of simplified protocol, which would:
    - Use more sources for reactivity information, such as:
        - Multiple MSDSs
        - Related tools, such as the NOAA Chemical Reactivity Worksheet
        - Standard references, such as Bretherick (8)
        - Vendor information

The efforts described in item 2 above are currently being undertaken by a separate working group (the Physical-Chemical Processing Subteam) within the RMR, led by Kenan Stevick of the Dow Chemical Company. Anyone interested in participating in the work of the RMR may contact Scott Berger for information on how to get involved.

Conclusion

This simplified approach provides a simple, rapid method for evaluation of, and solution for potential reactivity hazards in storage applications. Future enhancements will allow for greater ease of data input, and access to a wider variety of reference information.

References

1. U.S. Chemical Safety and Hazard Investigation Board, "Improving Reactive Hazard Management", Report 2001-01-H, Washington DC (October 2002).
2. Berger, Scott "*AIChE Initiatives to Promote Effective Management of Chemical Reactivity Hazards*", Process Safety Progress, Vol. 25, No. 3, pp174-179 (September 2006)
3. "Chemical Warehousing: The Storage of Packaged Dangerous Substances", HSG71, Health and Safety Executive of the UK, Third Edition (2003)
4. National Fire Protection Association (NFPA), Batterymarch Park, Quincy, MA, NFPA 30 "Flammable and Combustible Liquids Code" (2003)
5. National Fire Protection Association (NFPA), Batterymarch Park, Quincy, MA, NFPA 430 "Storage of Liquid and Solid Oxidizers" (2004)
6. National Fire Protection Association (NFPA), Batterymarch Park, Quincy, MA, NFPA 432 "Storage of Organic Peroxide Formulations" (2002)
7. National Fire Protection Association (NFPA), Batterymarch Park, Quincy, MA, NFPA 495 "Explosive Materials Code" (2006)
8. Urben, P.G. (ed.) 1999. Bretherick's Handbook of Reactive Chemical Hazards, Sixth Edition. 2 vols. Oxford, UK: Butterworth Heinemann. ISBN 0-7506-3605-X. Also available on CD-ROM as Bretherick's Reactive Chemical Hazards Database - Version 3.0. Oxford, UK: Butterworth-Heinemann. ISBN 0-7506-4342-0 and on-line by subscription [http://www.chemweb.com].

Spreading the word of the concept 'inherent safety' in a general industrial setting in the Dutch province of Zeeland

Mat J.M. Jongen, Anja Dijkman, Sander Zwanikken, Gerard I.J.M. Zwetsloot, Johan Gort

TNO Quality of Life, Business Unit Work and Employment, Hoofddorp, The Netherlands, www.tno.nl. Email: mat.jongen@tno.nl

ABSTRACT

Recent accidents in The Netherlands in different kinds of industries, like fire works storage, catering and energy industry, triggered the Dutch government to start a national program to enhance the enforcement of industrial safety at the regional and municipal level. Stimulated by this program the Province of Zeeland developed its own external safety program. It explains the role external safety in major hazard control and in the co-ordination of land use planning. The pressure for land has for long been large in the Netherlands and the chance of major accidents influencing neighboring land has been strictly regulated. Part of the program is a feasibility study on introducing the concept of inherently safer solutions to undustry and local authorities. TNO executed a communication and exploration project for introducing this concept to the industry of Zeeland and other involved actors like licensing authorities and the regional fire fighting organization. Part of it was a search for examples of successful implementation of the concept. Ultimate goal of the Province is to enhance solutions directed at decreasing the most prominent hazards at source in this part of the Netherlands. These hazards concern large chemical facilities, production sites with large quantities of dangerous goods, large cooling installations and propane or LPG tanks. This paper presents the outcome of the communication and exploration campaign about the concept of inherent safety.

Introduction

In 2000 and 2001 several major accidents in the Netherlands have challenged the way the Dutch government, at national, provincial and municipal level, deals with safety. The 2001 fire works explosion in the town of Enschede killed several dozens of inhabitants and destroyed a whole city block. Also in 2001, the café fire in Volendam killed more than a dozen, predominantly young visitors. These and other accidents prompted the discussion on how external safety is organized and enforced in the Netherlands. Several government programs have been initiated to improve industrial safety through different approaches. A large program of the ministry of Social Affairs and Employment has been directed at finding new approaches for industrial safety, one of which was promoting workers safety through tackling risks at source [1]. Another program of the ministry of Housing, Spatial Planning and Environment is directed at strengthening the organizational setting of external safety. In this program, the national government provides funding for provinces who want to strengthen their policy regarding external safety.

TNO has since long been active in promoting sustainable development with respect to a healthy work force and working conditions. Developing new insight and methods for helping companies in improving their safety performance and working conditions is part of the research program. Although inherent safety is a well know principle with specialists in process safety and in the risk management of nuclear power plants, it is not commonly known in other branches of industry [2,3,4,5]. The province of Zeeland invited TNO in the fall of 2006 to explore the feasibility of introducing the concept of inherent safety to the relevant industries and authorities in the province. Which, if successful, should lead to a long lasting search for opportunities within companies and between companies and authorities for the application of inherent safer solutions at the production sites in Zeeland.

External safety in the province of Zeeland

The province has a so-called risk chart with all companies that pose a certain risk to their environment (www.zeeland.nl, in Dutch). In total, some 200 companies appear in this chart. The province is the fore-runner province in executing the national program for advancing the management of external safety in the Netherlands. It has developed a well defined policy of strengthening the infrastructure and organizational aspect of external safety. The implementation of laws dealing with external safety of production sites BRZO (the Dutch implementation of the European Seveso II directive), spatial planning of production sites, directives for transport of dangerous goods and the licensing of dangerous production sites has been adapted for stricter risk management. The basis for the projects in Zeeland dealing with external safety is the provincial policy document 'Risico's Inzicht' ('Risks Insight '). One of the program elements was to explore possibilities for decreasing risks by tackling them at source i.e. trying to reduce or eliminate risks in stead of trying to merely control the existing risks. The above mentioned sites consist mostly of chemical process industry, companies with large stocks of dangerous chemicals, with cooling installations containing dangerous cooling agents, and companies with large propane or LPG tanks. The provincial administration would like to reduce these risks as much as possible, however, with the explicit message that it does not want to decrease the economic strength of the province.

The communication and exploration project

The provincial program for decreasing external risks runs from 2006 until 2010. One of the themes in the program is an attempt to introduce the concept of inherent safety to a wider public, especially the companies on the provincial risk chart and the authorities that license companies to build or modify their plants. Aim is to identify opportunities for decreasing or even eliminating (external) risks and to find, if possible, examples of successful implementation of the concept. Part of the exploration was a discussion on the possible role of provincial and municipal licensing authorities in promoting the principle of inherent safety during the licensing process. Not as a formal part of this process, but as a way to start an open discussion on safer alternatives for the company's production process.

The concept has been presented by TNO in several meetings of company networks in the province, i.e. a network focused on sustainable production and its working group on safety. In addition, all relevant companies on the risk chart of the province were invited to a starting meeting where the concept was explained and companies were stimulated to present examples of their own successful projects in making processes inherently safer. License officers of the province of Zeeland and the 13 municipalities have been introduced to the concept of inherent safety by interviews and a presentation in a meeting. This communication and exploration project ran from September 2006 until January 2007.

Some 16 companies participated in the first starting meeting. Eight of them showed interest in participating and staying informed on the subject. Several companies came up with projects they wanted to start that could improve safety or decrease environmental risks. One company presented a successful example of inherently safer working: a large propane tank for the production of steam for heating purposes had been replaced by a connection to the Dutch natural gas network resulting in a substantial decrease of the 10^{-6} safety contour and additional energy savings through improved process efficiency. The starting meeting was reported in an article in the monthly entrepreneurs issue of the local news paper, reaching some 80 000 subscribers. Discussion in the working group on safety yielded points of application were the principle of inherent safety may be incorporated in a company's safety management, i.e. the evaluation of incidents, the modification of existing plants and the construction of new installations.

The interviews with the license officers showed positive responses and a willingness to learn more about the concept and ways to implement it in daily practice. Additional training appeared to be required. The interviews yielded a substantial number of suggestions for putting the concept of inherent safety on the agenda in the process of granting licenses. All in all this three months communication and exploration project has yielded a quite satisfying response, keeping in mind that the concept of inherent safety was not commonly known in the companies and unknown with most of the license officers. At the time of writing of this paper a proposal for a follow-up was in discussion in the provincial co-ordination group for external safety.

Discussion

As far as we know, this is the first time that regional or local authorities in the Netherlands try to introduce the concept of inherent safety to a wider public and to its own personnel involved in providing licenses to local companies. This offers a unique opportunity for starting a trend that will stimulate companies to think differently about safety with a focus on innovation that will reduce the risks for the company and its environment. Putting the concept of inherently safer production on the agenda of all involved parties in a particular region requires a series of implementation steps that - in general - consists of agenda setting and creating awareness about the concept. Followed by strengthening willingness with all involved actors, then providing them with knowledge and tools to implement the concept, and finally finding opportunities for actual execution of projects that will improve industrial safety. This requires a long term commitment of the companies involved as well as of provincial and municipal authorities. The current project focused on the first steps of this implementation through agenda setting and communication. Communication was directed at organizing meetings with the most important actors, safety managers from the companies and licensing officers from the local authorities and the fire fighting organization. This led to many discussions about the concept inherent safety and how to implement it in daily practice. This has certainly contributed to enhancing the understanding of the concept of inherent safety.

The project was a feasibility study for a further introduction of the concept. In fact, the first step in the above-mentioned implementation plan. During the meetings it became clear that the concept inherent safety was not very well known and needs considerable explanation and discussion. The positive result of the exploration and communication activities was that half of the companies present at the starting meeting showed interest in the concept and even came up with ideas for future projects. Regarding the license officers, we noted the willingness to continue the exploration of the concept. One of the options for further exploration was discussing the concept and the opportunities it offers with companies in the informal preparation phase of the licensing process.

Application of the principle of operating inherently safer needs a window of opportunity. Changes to processes or installations cannot always be applied directly. In order to find all possible improvements in safety, inherent safety needs to be on the agenda of companies and authorities in a structural way for a prolonged period of time. Organizing this would be the next step in the implementation. Linking the concept to the ongoing, structural initiatives for a sustainable industry in the province and to the normal process of providing licenses to companies is the way to go. Tools for the implementation within companies and the licensing process will be needed and can be developed in the coming year. TNO will also continue collecting examples of successful implementation projects that can help explain the concept of inherent safety and will stimulate other companies to explore their own opportunities for safer production.

References

1. Sander, Zwanikken LJ, Mat, Jongen JM, Jan, Popma R, Raphaël, Gallis R, Gerard, Zwetsloot IJM. Tackling occupational hazards at source. A search for innovative methods to implement source-directed strategies. Working on Safety Conference, Zeewolde, The Netherlands, september 2006.
2. Robert, Bollinger E, et al. Inherently safer chemical processes. A life cycle approach. A CCPS Concept Book, American Institute of chemical engineers. Center for Chemical Process Safety, New York, 1996.
3. Trevor Kletz. Process Plants: a handbook for inherently safer design. Taylor & Francis. Philadelphia, USA, 1998.
4. Nicholas, Ashford A, Gerard Zwetsloot, IJM. Encouraging inherently safer production in European firms: a report from the filed. J. Hazardous Materials 78 (2000) 123-144.
5. Gerard, Zwetsloot, IJM, Nicholas, Ashford A. The feasibility of encouraging inherently safer production in industrial firms. Safety Science, 41 (2003), 219-240.

Regulating Inherent Safety
Contra Costa County's Industrial Safety Ordinance

Authors: Randall L. Sawyer, Cho Nai Cheung, Michael Dossey, Don Nixon, Nicole Alaniz, and Habib Amin

The Contra Costa County Board of Supervisors, California, approved the County's Industrial Safety Ordinance in December 1998; it became effective on January 17, 1999. The ordinance was passed to address the causes of the accidents that occurred at Contra Costa County's Petroleum Refineries and Chemical Plants in the late 1980's and 1990's. The Industrial Safety Ordinance expands on the California Accidental Release Prevention Program, which is more stringent than the Federal Risk Management Program. One element of the Industrial Safety Ordinance requires regulated stationary sources to implement inherently safer systems wherever feasible. By requiring stationary sources to consider inherent safety, a reduction of hazards has been accomplished. Since the Industrial Safety Ordinance was implemented, the number of major chemical accidents or releases has decreased.

This paper discusses what has worked and what has not worked regulating the covered facilities. The paper also shows how inherent safety is being considered and implemented by the covered facilities.

Background

Contra Costa County is across the San Francisco Bay from San Francisco and is the home to four petroleum refineries and many small to medium sized chemical plants. More than one million people reside in the County. There are nineteen cities in the County including Richmond, Concord, Antioch, Pittsburg, and Walnut Creek.

The California Legislature approved the Risk Management and Prevention Program in 1986, one of the first in the country. California replaced the Risk Management and Prevention Program in 1997 with the California Accidental Release Prevention Program, which is more stringent then the Federal Risk Management Program. Contra Costa Health Services Hazardous Materials Programs has been responsible for implementing the state programs in Contra Costa County.

Risk Management and Prevention Program

The Risk Management and Prevention Program required facilities that used more than a threshold quantity of an "Acutely Hazardous Material" to submit a Risk Management and Prevention Plan to the local Administering Agency, when requested to do so by the Administering Agency. This program also required facilities to reduce the possibility of accidental releases of toxic chemicals that could impact the public. The statute required facilities to perform Hazard and Operability Studies to assess the risk of accidents and reduce it with additional safeguards when the risk was found to be too high.

Contra Costa County started implementing the Risk Management and Prevention Programs requirements in 1989 and 1990 and the result was a 34 percent decrease on the amount of Acutely Hazardous Materials in the County. Figure 1 shows a breakdown of some of the reduction for specific Acutely Hazardous Materials from the largest quantity of the chemical handled in the County between 1990 and 1994 and the quantity in 1994.

The businesses in Contra Costa County started to consider and apply inherently safer changes to their operations as a result of implementation of the Risk Management and Prevention Program. There have been continued and dramatic reductions in the inventories of Acutely Hazardous Materials in the County. All of the wastewater treatment plants have eliminated the use of sulfur dioxide, chlorine, and ammonia through substitution of a different chemical or using ultraviolet light process. All but three of the water treatment plants have eliminated chlorine and ammonia through substitution. All of the refineries have eliminated chlorine use through substitution. Most

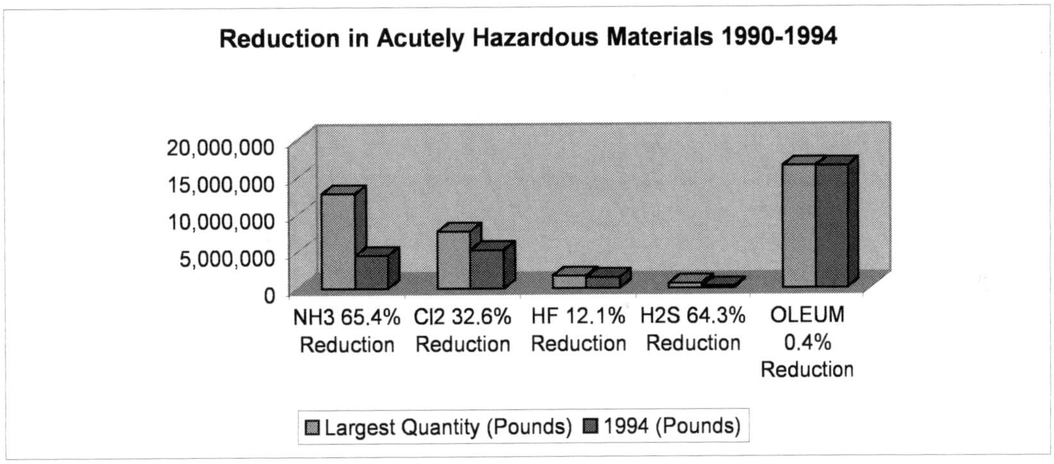

Figure 1

of the selective catalytic reduction process for control of nitrogen oxides has substituted ammonium hydroxide for ammonia. Since 1994, all but two of the large public swimming pools have substituted sodium or calcium hypochlorite or a chlorine generator for chlorine. A spent sulfuric acid regeneration plant reduced the amount of oleum stored by 90% and also reduced the concentration from 35% to 25% by changing the way they produce an electronic grade sulfuric acid. A refinery reduced the amount of ammonia storage from 8.9 million pounds to 1.7 million pounds and also moved the storage location approximately one mile further away from the community then where it was being stored.

Industrial Safety Ordinance

The Risk Management and Prevention Program was replaced by the California Accidental Release Prevention Program in January 1997. The County Board of Supervisors' concern about accidents that occurred in the County in the late 1980s and 1990s (see Table 1) led to the passage of an Industrial Safety Ordinance in 1998. The ordinance requires more prevention elements and more public participation for refineries and chemical plants than either the Federal Risk Management or the California Accidental Release Prevention Programs. The ordinance expands the application of human factors beyond the Process Hazard Analysis and requires facilities to do the following: include a Root Cause Analysis as part of Incident Investigations; review and manage Contra Costa County Major Accidents at Chemical Plants and Petroleum Refineries changes to responsibilities and organization of the facility; apply Inherently Safety Systems for new and existing process; perform a Security and Vulnerability Analysis; and complete periodic Safety Culture Assessments. Significantly, the entire facility is covered under the Ordinance, not just the processes with a regulated substance above a threshold quantity.

Inherently Safer Systems

Industrial Safety Ordinance §450-8.014(g) defines inherently safer systems (ISS) as:

> *"'Inherently safer systems' means "inherently safer design strategies" as discussed in the latest edition of the Center for Chemical Process Safety Publication Inherently Safer Chemical Processes, and means feasible alternative equipment, processes, materials, lay-outs, and procedures meant to eliminate, minimize, or reduce the risk of a major chemical accident or release by modifying a process rather than adding external layers of protection. Examples include, but are not limited to, substitution of materials with lower vapor pressure, lower flammability, or lower toxicity; isolation of hazardous processes; and use of processes which operate at lower temperatures and/or pressures."*

Industrial Safety Ordinance §450-8.016(d)(3) states when inherently safer systems are to be considered as follows:

> *"For all covered processes, the stationary source shall consider the use of inherently safer systems in the development and analysis of mitigation items resulting from a process hazard analysis and in the design and review of new processes and facilities. The stationary source shall select and implement inherently safer systems to the greatest extent feasible. If a stationary source concludes that an inherently safer system is not feasible, the basis for this conclusion shall be documented in meaningful detail."*

Initial Implementation of Inherently Safer Systems

Regulated stationary sources are required to implement inherently safer systems to the greatest extent feasible for new processes and facilities and also in the development and analysis of mitigation items from process hazard analysis. Regulated stationary sources are to follow guidance provided by the Center for Chemical Process Safety's (CCPS) Inherently Safer Chemical Processes book.

TABLE 1

Summaries of major chemical or refinery accidents in Contra Costa County in 1990's

Date Accident Occurred	Accident Description	Onsite Impact	Offsite Impact
March 25, 1999	Petroleum fire that required the sounding of sirens. A valve stem blew out causing a release of hydrocarbons and hydrogen sulfide that ignited, resulting in a major fire.	Shutdown a major part of the refinery. Two people slightly injured fighting the fire.	Smoke was visible over much of the North Bay Area. Disruption in traffic, sirens sounded, people were asked to shelter-in-place.
February 23, 1999	Flash fire at a crude unit killing four people and seriously injuring another.	Shutdown of the crude unit and later the whole refinery. Four people died with another seriously injured.	None.
April 23, 1997	An upset in the distillation unit sent hydrocarbons to the sulfur recovery units. The sulfur recovery units released hydrogen sulfide.	Shutdown parts of the refinery until the problem was found and the different units were stabilized.	People were asked to shelter-in-place.

Summaries of major chemical or refinery accidents in Contra Costa County in 1990's

Date Accident Occurred	Accident Description	Onsite Impact	Offsite Impact
January 27, 1997	A hydrocracker had a runaway reaction that raised the temperature and pressure of the outlet gases. An outlet pipe failed and caused an explosion and fire. One worker was killed and 46 people sought medical attention.	One worker was killed and 46 other sought medical attention. Many of the people seeking medical attention were injured while running away from the explosion and fire. The outlet piping and some of the equipment and instrumentation was damaged. The hydrocracker unit was shutdown until the unit could be repaired.	There was little or no offsite impact from this explosion and fire. People were not asked to take any protective action, such as sheltering place.
May, 1996	A hot coke drum was dumped and a fire was initiated.	Fire damage to the coker unit. The coker unit was run at partial capacity until repairs could be made.	People were asked to shelter-in-place. A few people sought health care assistance complaining of the smoke and odors.
April 1, 1996	Fire at their cracked gas unit. The fire lasted more than three hours.	Fire damage was extensive onsite. The unit was shutdown until repairs could be made.	People were asked to shelter-in-place downwind of the refinery. The smoke from the fire impacted the area downwind.

Summaries of major chemical or refinery accidents in Contra Costa County in 1990's

Date Accident Occurred	Accident Description	Onsite Impact	Offsite Impact
February 1, 1996	Piping failed because of corrosion and a release of hydrogen occurred with an explosion. The fire lasted approximately 10 minutes.	Shutdown of the hydrogen unit.	Noise of the explosion was heard offsite for well over a mile. Some minor damage offsite.
June 16, 1995	Petroleum tank fire that continued to burn off and on for approximately one week.	Response by the facility along with local fire departments.	Smoke impacted the surrounding communities. The communities were asked to shelter-in-place. Odors lingered in the communities for a week after the fire. The refinery put up approximately 100 people in motels until the incident was 100% secure.

Summaries of major chemical or refinery accidents in Contra Costa County in 1990's

Date Accident Occurred	Accident Description	Onsite Impact	Offsite Impact
August 22 - September 6, 1994	A column used to strip carbon dioxide developed a leak. The column continued to operate for a 16-day period. The hole in the column became larger where a solution of Catacarb® was atomized and released. Estimates of between 80-225 tons of Catacarb® was released.	The Catacarb® affected the health of many of the workers downwind of the column. At the end, the hydrocracker processing unit was shutdown.	The release impacted the surrounding communities. Many homes and cars were cleaned to remove the Catacarb® solution. A health clinic setup by the refinery that treated more than 1,200 people over the next year.
July 26, 1993	Overheating of a tank car of 35% oleum led to a rupture disk relieving that released between 4 and 8 tons of sulfur trioxide over a three hour and forty-five minute period. The sulfur trioxide formed a sulfuric acid cloud.	The loss of the contents of a tank car of 35% oleum and the shutting down of their processing.	Over 22,000 people sought medical assistance, with 15-22 people staying in the hospital overnight. People were asked to shelter-in-place downwind of the release.

Summaries of major chemical or refinery accidents in Contra Costa County in 1990's

Date Accident Occurred	Accident Description	Onsite Impact	Offsite Impact
May, 1992	Sulfuric acid mixed with hydrocarbon spilled. The hydrocarbons ignited. The fire was difficult to put out because the sulfuric acid would break down the foam used by the fire fighters. One employee was killed and another was seriously injured.	Fire damaged some of the equipment onsite. One person died and another was seriously injured	Smoke went offsite and impacted the community. The smoke also included the degradation products of the sulfuric acid (e.g., sulfur trioxide, sulfur dioxide). Shelter-in-place was called for the community downwind of the fire.

Regulated stationary sources are required to implement the highest level of inherent safety strategies. The strategy levels of implementation are at follows:

- Inherent – Eliminating the hazard by using materials and process conditions which are non-hazardous; e.g., substituting water for a flammable solvent.
- Passive – Minimizing the hazard by process and equipment design features that reduce either the frequency or consequence of the hazard without the active functioning of any device; e.g., the use of equipment rated for higher pressure.
- Active – Using controls, safety interlocks, and emergency shutdown systems to detect and correct process deviations; e.g., a pump that is shut off by a high-level switch in the downstream tank when the tank is 90% full. These systems are commonly referred to as engineering controls.
- Procedural – Using operating procedures, administrative checks, emergency response, and other management approaches to prevent incidents or to minimize the effects of an incident; e.g., hot-work procedures and permits. These approaches are commonly referred to as administrative controls.

Regulated stationary sources are to apply the different approaches for implementing inherent safety as follows:

- Minimization – Use smaller quantities of hazardous substances (also called *Intensification*)
- Substitute – Replace a material with a less hazardous substance
- Moderate – Use less hazardous conditions, a less hazardous form of a material, or facilities that minimize the impact of release of hazardous material or energy (also called *Attenuation* or *Limitation of Effects*)
- Simplify – Design facilities that eliminate unnecessary complexity and make operating errors less likely, and that are forgiving of errors that are made (also called *Error Tolerance*)

Contra Costa Health Services (CCHS) Hazardous Materials Programs) is required to audit and inspect each of the regulated stationary sources at least once every three years. As part of the audit, CCHS ensures that the regulated stationary sources are implementing inherently safer systems to the greatest extent feasible. During the first round of audits in 1999, CCHS found that when regulated stationary sources tried to apply inherently safer systems to Process Hazardous Analyses mitigations, the distinction between inherent safety and risk reduction was lost. CCHS found that any risk reduction implemented was considered inherent safety. Regulated stationary sources did use the inherent safety strategies in considering risk reduction, which was an improvement in the consideration of what risk reduction methods to use. Regulated stationary sources followed overly broad interpretations of inherent safety in the CCPS book Inherently Safer Chemical Processes A Life Cycle Approach instead of working to implement the highest level of inherent safety (i.e., moving up the different levels from procedural to active to passive to inherent).

Inherently Safer Systems Guidance

It became clear to CCHS that additional guidance was needed on when and how to consider inherent safety. CCHS worked with the regulated stationary sources and developed this guidance, which was broken down into seven sections. The first section addresses new covered processes; the second addresses existing processes; the third addresses mitigations resulting from Process Hazard Analysis (PHA); the fourth defines feasibility; the fifth addresses recommendations from process hazard analyses; the sixth addresses Inherently Safer System Reports; and the seventh contains definitions. The first four sections are discussed below.

New Process or Facilities

Inherent safety must to be considered for new processes at different phases in the development of a project. Inherently Safer Systems should be reviewed early in the development phase of a new covered process and then reviewed throughout the different project design phases. The objectives for an inherent safety review are to employ synergistic teams. Composition of the review teams will vary at different phases of the development cycle of the project and with the nature of the process. In product development and design scope, the team may comprise of chemist, process design engineers, industrial hygienist, safety engineer, environmental, and control engineer etc. In the hazard and operations review phase, the team may include operations and maintenance personnel as well. A stationary source may choose not to use an ISS review team at a particular phase. If this does occur, the stationary source should document the reasons for not using a team for performing an ISS review during this particular phase. A multi-disciplined group should be used throughout the various phases of development, implementation, and operation. The team is purpose is to:

- Understand the hazards
- Find ways to reduce or eliminate the hazards

> *"The first major objective for the inherent safety review is the development of a good understanding of the hazards involved in the process. Reducing and eliminating hazards and their associated risks is the second major objective. Applying inherent safety principles early in the product/process development effort provides the greatest opportunity to achieve the objectives of the inherent safety review process for the project at hand."*[1]

The stationary source should use a review process for new processes that includes an Inherently Safer Systems Review at different phases of the design process, such as the following phases when applicable:

- During the chemistry forming (synthesis) phase for product/process research and development to focus on the chemistry and process
- During the facilities design scoping and development prior to completion of the design basis to focus on equipment and configuration

---

[1] CCPS, Inherently Safer Chemical Processes, A Life Cycle Approach, 1996

- During the basic design phase of the project

Existing Processes

The Industrial Safety Ordinance requires that stationary sources consider hazards as part of the process hazard analyses.

> *"The process hazard analysis shall be appropriate to the complexity of the Covered Process and shall identify, evaluate, and control the hazards involved in the Covered Process. The process hazard analysis shall address: the hazards of the process; the identification of any previous incident which had a likely potential for catastrophic consequences; engineering and administrative controls applicable to the hazards and their interrelationships such as appropriate application of detection methodologies to provide early warning of releases."* Chapter 450-8 §450-8.016(d) (1)

Stationary sources should use one of the following methods to ensure that inherently safer systems (inherent and passive categories) are considered and documented for the covered processes:

- An independent inherently safer system analysis that is done in addition to a Process Hazard Analysis (PHA)[2]. These analyses should review the covered processes for ways to eliminate or reduce hazards that are present in the covered process. This may be achieved by using a checklist or guideword analysis that incorporates inherently safer systems. If the stationary source decides to use some other inherently safer systems checklist or other methods to evaluate inherently safer systems, Contra Costa Health Services must approve these prior to their use.

- An inherently safer system analysis that is incorporated into the existing process hazard analysis review process. This would require that each covered process in its entirety has an initial inherently safer systems analysis conducted. (Incorporating inherently safer systems into a revalidated process hazard analysis may not be sufficient to satisfy the initial inherently safer system review if the whole process is not evaluated.) This may be achieved using a checklist or guideword that incorporates inherently safer systems considerations into a Hazard and Operability Study where a Major Chemical Accident or Release could reasonably occur. These analyses would review the covered processes for ways to eliminate or reduce hazards as well as risks that are present in the covered process.

---

[2] If the stationary source decides to do an independent inherently safer systems analysis, CCHS suggests that this be done in conjunction with the process hazard analysis, but it may be appropriate for a stationary source to perform an inherently safer systems analysis that is done at a different time than the process hazard analysis. Either approach meets the guidance from this document, as long as the inherently safer systems analysis is revalidated at least once every five years.

Process Hazard Analysis Recommendations and Mitigations

The concepts as addressed in the CCPS book Inherently Safer Chemical Processes A Life Cycle Approach for looking at all four categories of risk reduction (Inherent, Passive, Active, and Procedural) should be used in the development and mitigation of recommendations from process hazard analysis as well as for considering Human Factors. Chapter 4 of the CCPS book Inherently Safer Chemical Processes A Life Cycle Approach discusses many inherently safer system strategies that can be incorporated in the development of mitigations to address the recommendations from process hazard analysis.

The stationary sources should provide guidance to personnel responsible for developing and analyzing recommendations and mitigation items resulting from the unit PHA. The guidance should include the concepts of inherently safer systems including:

- The different categories of risk reductions
- Moving up the different levels from procedural to active to passive to inherent
- Approaches to apply inherently safer systems including minimization, substitution, moderation, and simplification

Feasibility

One of the areas of greatest concerns to the regulated stationary sources is what would be considered feasible. The fourth section of the guidance defines feasibility. To assist in the determination of feasibility, Contra Costa Health Services is using a modification of the following guidance from OSHA and the U. S. EPA. (Federal OSHA provided guidance for justifiably declining recommendations from incident investigations in the September 1994, OSHA Instruction CPL 2-2.45A CH-1. These criteria have since been applied to recommendations formulated during Process Hazard Analysis.) The version used by Contra Costa Health Services uses the following criteria for infeasibility:

- The analysis upon which the recommendations are based contains factual errors.
- The recommendation is not necessary, i.e., the safeguards may be inadequate, but the consequences are operational or the consequence or severity of the scenario would not result in a Major Chemical Accident or Release.
- An alternative inherently safer system would provide a sufficient level of hazard reduction (NOTE: Implementing only one option to address identified hazards may not be adequate to address the greatest hazard reduction or elimination. However, it is not necessary to implement more than one inherently safer system if the implementation of a second inherently safer system does not add any significant hazard reduction or has been documented as infeasible.)
- The recommendation is in conflict with existing federal, state, or local laws.
- The recommendation is in conflict with Recognized and Generally Accepted Good Engineering Practices (RAGAGEP).
- The recommendation is economically impractical, such that the process unit can no longer be financially operated. This can include the following factors:
    - Capital investment
    - Product quality
    - Total direct manufacturing costs

  - Operability of the plant
  - Demolition and future cleanup and disposal cost
- The recommendation would have a negative social impact such that the project should not be implemented. Some examples of social impact are a visual or noise impact on the community that is not acceptable or increased the traffic congestion.
- The recommendation may violate a license agreement and the license agreement cannot be modified and must remain in effect.
- The recommendation may decrease the hazard, but would increase the overall risk.
- An alternative measure would provide more risk reduction than the inherently safer system.

If the inherently safer system recommended is determined not to be implemented because it will create more risk, or if other modifications that are not inherently safer system are made such that the overall risk is less than if the inherently safer system were implemented, the stationary source will need to document how this determination was made. A qualitative risk assessment/analysis could be used as a basis for the determination. The facility needs to document how they determined the qualitative severity and likelihood for the existing or modified conditions and for the conditions if the ISS is implemented. If the qualitative risk analysis shows the same level of risk, then a quantitative risk assessment/analysis should be performed to compare the risk of the existing or modified conditions to the risk if the ISS is implemented. The documentation should include the background information that was used to do the comparison of the existing or modified conditions to the conditions if the ISS is implemented. Another method may be used by the stationary source, such as a weighted scoring decision matrix as shown on page 23 of CCPS book Inherently Safer Chemical Process A Life Cycle Approach, if the method is approved by Contra Costa Health Services prior to the use of the method.

County Ordinance Code Chapter 450-8, §450-8.016(d)(3) requires the following: "This documentation shall include: (1) sufficient evidence to demonstrate to the County's satisfaction that implementing this inherently safer system is impractical, and (2) the reason for this conclusion." The documentation should include the applicable background information, calculations, and the reasons why an inherently safer system was not implemented. If there is any concern that the reasons for not implementing an inherently safer system would not satisfy Contra Costa Health Services, the stationary source should consult with Contra Costa Health Services to determine if the justification is satisfactory. The stationary source should then receive in writing from Contra Costa Health Services their decision and how they came to their decision, including background information, calculations, and alternatives considered.

Implementing Inherently Safer Systems

The regulated stationary sources have applied inherently safer systems considerations to their existing and new processes since the guidance has been completed; include applying independent ISS analyses and incorporating the ISS evaluation as part of their existing process hazard analyses. Also, the regulated stationary sources have applied the guidance for any new processes or facilities, and in the mitigations of recommendations from process hazard analyses. Applying inherently safer systems to process hazard analyses mitigations allows a regulated stationary source to use the thought process for implementing inherent safety, including ways of eliminating or reducing the hazard. Table 2 shows some of the ISS that have been implemented by the regulated stationary sources.

Effect of the Industrial Safety Ordinance including Inherently Safer Systems

Figure 2 indicates that since the Industrial Safety Ordinance was implemented, the number of major chemical accidents or releases has decreased. A major portion of the County's ordinance is inherently safer systems. Further analysis and audits over a longer period of time will be needed to determine how much of the reduction may be attributed to the requirement to implement the above described ISS process.

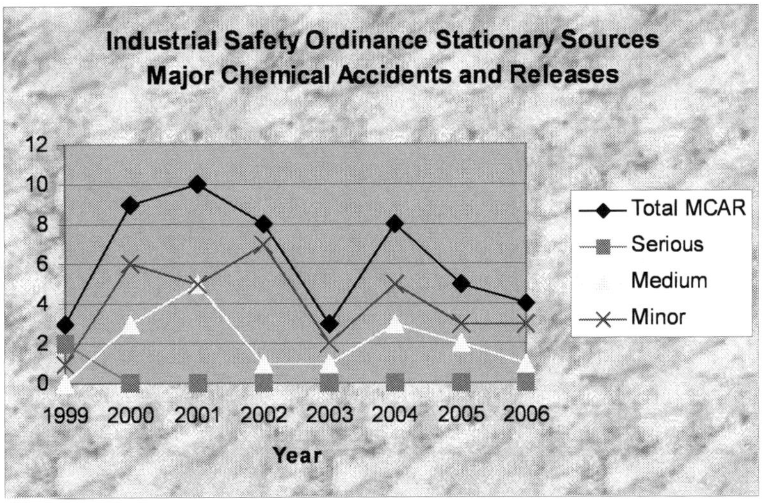

Figure 2

Major Chemical Accidents and Releases are defined in the Industrial Safety Ordinance. The types of Major Chemical Accidents can be broken down further as shown in Figure 2 and defined as follows:

- Serious – A fatality, serious injuries, or major onsite and/or offsite damage occurred
- Medium – An impact to the community occurred, or if the situation was slightly different the accident may have been considered major, or there is a recurring type of incident at that facility
- Minor – A release where there were no injuries or minor injuries, the release had no impact or slight impact to the community, or there was no damage or minor onsite damage

TABLE 2

Summary of ISS and Risk Reduction Implemented in Contra Costa County

		2006	2005	2004	2003
Inherent:	Minimization	· Reduction of inventory (4)	· Reduction of inventory (11)	· Reduction of inventory (11) · Eliminated a unit and regulated material (1) · Incorporated existing equipment into a new unit and reduced piping (1)	· Eliminated hazard (2) · Reduction of inventory (9) · Process re-designed to eliminate need for a second tower (1)
	Substitute		· Process changed using less hazardous chemical (1)		
	Moderate		· Reduced the hazard by changing hazard conditions or form (4)		· Decreased concentration (1) · Less hazard process conditions or form (3)
	Simplify	· Removed out of service equipment or dead leg (8) · Simplified process (1)	· Simplified the process or eliminated equipment (5)	· Revised equipment design features (5) · Simplified the process (5) · Eliminated cross contamination connections (3)	· Eliminated equipment that could cause hazard (2)
Passive:	Minimization	· Revised equipment design features (1)			
	Substitute				

	Moderate	· Reduced the potential of a hazard by eliminating the hazard or moving to an alternate location (2)			
	Simplify	· Improved capacity, temperature, and/ or overpressure design rating of equipment (3) · Reduced potential or frequency of a hazard by changing design features and/or installation of blinds (8)	· Reduced potential overpressure hazard by equipment elimination (8) · Reduced potential of a hazard by design features, re-routing piping, and increased support (7) · Improved temperature and pressure design rating of equipment (11) · Reduced the potential of a hazard by eliminating liquid accumulation points or vibration stress (5)	· Revised equipment design features (2) · Improved temperature design rating of equipment (2) · Reduced the frequency of a hazard (5)	· Simplified the process or piping (4) · Reduced potential of a hazard by design features, heating media and equipment location (8) · Improved overpressure design rating of equipment (3) · Eliminated hazard to downstream unit without active device by removing contaminants from system (1)

Conclusions

Contra Costa Health Services has found that risk management and accident prevention regulations alone have reduced the amount of acutely hazardous materials that are handled in Contra Costa County by 34 percent within the first four years of requiring such programs. CCHS has also found that it was necessary to provide detailed guidance to ensure consistent and thorough implementation of inherently safer concepts by the regulated stationary sources. By requiring stationary sources to consider inherent safety, a reduction of hazards has been accomplished. Is it necessary for regulations to be developed to ensure application OF inherently safer systems analyses and a reduction of hazards? Requiring stationary sources to consider inherent safety been effective in Contra Costa County in reducing the hazards that exist at regulated stationary sources. Industry and the process safety community have the responsibility and obligation to share knowledge and encourage the use of inherent safety beyond what is required.

CCPS 22nd Annual Conference on Case Histories:
Advancing Process Safety Through Design and Operations
April 22-26, 2007
Houston, TX

Inherently Safer Chemical Processes - 2nd Edition

David A. Moore
Mike Hazzan
Marty Rose
David Heller
AcuTech Consulting Group
2001 North Beauregard Street
Alexandria, VA 22311

Dennis C. Hendershot
Consultant
534 Norris Drive
Furlong, PA 18925

Arthur. M. Dowell, III, PE
Rohm and Haas Company
6519 State Hwy 225
Deer Park, TX 77536

Abstract

The classic reference from CCPS "Inherently Safer Chemical Processes, A Life Cycle Approach", 1st Edition, 1996[1], was updated in 2006-7. The goal of this book is to influence the future state of chemical process evolution by illustrating and emphasizing the merits of integrating process research, development, and design into a comprehensive process that balances safety, capital, and environmental concerns throughout the life cycle of the process. The authors hope that this book will influence the next generation of engineers and chemists as well as current practitioners and managers in the field of chemical processing.

Lessons learned since the first edition of the CCPS® inherent safety concept book in 1996, combined with the fact that inherently safer design (ISD) is becoming more widely accepted prompted CCPS® to update the book. Also, since 1996, several jurisdictions have mandated consideration of inherently safer design for certain facilities, and such requirements have been proposed at the Federal level in the United States. In particular, there is a need for more guidance, especially in practical, step-wise approaches to conduct inherently safer studies. This edition builds on the first edition with the same philosophy but clarifies the

[1] Center for Chemical Process Safety (CCPS) (1996). *Inherently Safer Chemical Processes - A Lifecycle Approach.* New York: American Institute of Chemical Engineers.

concept with recent research and thoughts of practitioners, added examples, added more industry methods, security issues and a discussion on regulatory issues. A key feature of the new book is the clarification of the meaning of inherently safer design by setting 1st and 2nd 'orders' of inherently safer design. This is distinguished from layers of protection and other traditional process safety concepts.

Numerous added examples and checklists make this a very practical guide and a good edition to the library of anyone involved in process risk management.

Introduction

For over 30 years the American Institute of Chemical Engineers (AIChE) has been involved with process safety and loss control issues in the chemical, petrochemical, hydrocarbon process and related industries and facilities. AIChE publications and symposia are information resources for the chemical engineering and other professions on the causes of process incidents and the means of preventing their occurrences and mitigating their consequences.

The Center for Chemical Process Safety (CCPS), a Directorate of the AIChE, was established in 1985 to develop and disseminate technical information for use in the prevention of major chemical process incidents. With the support and direction of the CCPS Advisory and Managing Boards, a multifaceted program was established to address the need for Process Safety Management systems to reduce potential exposures to the public, the environment, personnel, and facilities. This program involves the development and publication of *Guidelines* relating to specific areas of Process Safety Management; organizing, convening and conducting seminars, symposia, training programs, and meetings on process safety-related matters; and cooperation with other organizations, both internationally and domestically, to promote process safety. CCPS has extended its publication program to include a "Concept Series" of books. These books are focused on more specific topics and are intended to complement the longer books in the *Guidelines* series. CCPS activities are supported by funding and professional expertise of numerous supporting organizations. Several government agencies and academic institutions also participate in CCPS endeavors.

In 1989, CCPS published the *Guidelines for Technical Management of Chemical Process Safety*, which presented a model for Process Safety Management characterized by twelve distinct, essential and interrelated elements. The Foreword to that book stated:

For the first time, all the essential elements and components of a model of a technical management program in chemical process safety have been assembled in one document. We believe the Guidelines provide the umbrella under which all other CCPS Technical Guidelines will be promulgated.

This "Concept Series" book, *Inherently Safer Chemical Processes: A Life Cycle Approach* supports many of the twelve Key Elements of Process Safety, as identified in the *Guidelines for Technical Management of Chemical Process Safety*, including Capital Project Review and Design Procedures, Process Risk Management, Human Factors, Audits and Corrective Action, and Enhancement of Process Safety Knowledge. The purpose of this "Concept Series" book is to demonstrate the application of inherently safer strategies throughout all the stages of the chemical process life cycle.

The Center for Chemical Process Safety (CCPS) of the American Institute of Chemical Engineers approved a 2nd Edition of the book in 2005. The intent of the committee was to use the existing 1996 edition of the book as a starting point, to update and modify the existing text where appropriate based on progress in the field since 1996, and to incorporate new material identified by the Inherently Safer Design Subcommittee (ISDS) during its initial deliberations on the project. Key objectives of the 2nd edition were to primarily focus on updating this text where appropriate based on reader feedback and progress in the area of inherently safer design since 1996, and also on incorporating new material identified by the ISDS. The project is titled Inherently Safer Design, and the book is intended to describe the current state of the art in the application of inherently safer design to chemical technology.

Inherent Safety has been well received by industry, but there has been significant advancement in the concept of inherently safer design over the last 10 years. The concept has received increased attention from industry, academia, government and regulatory authorities, and even the news media and general public. The existing 1996 book is frequently cited as an authoritative source on inherently safer design, and it is important that this book be updated to reflect current knowledge on the subject. The 2nd edition of the book is expected to be highly influential in the continuing discussion of the inherently safer design philosophy.

Objectives, Intended Audience and Scope of The Book

The goal of the updated inherent safety concept book is to influence the future state of chemical process evolution by illustrating and emphasizing the merits of integrating process research, development, and design into a comprehensive process that balances safety, capital, and environmental concerns throughout the life cycle of the process. The authors hope that this book will influence the next generation of engineers and chemists as well as current practitioners and managers in the field of chemical processing.

Lessons learned since the first edition of the CCPS® inherent safety concept book[2] in 1996, combined with the fact that inherently safer design (ISD) is becoming more widely accepted prompted CCPS® to update the book. Also, since 1996, several jurisdictions have mandated consideration of inherently safer design for certain facilities, and such requirements have been proposed at the Federal level in the United States. In particular, there is a need for more guidance, especially in practical, step-wise approaches to conduct inherently safer studies. This edition builds on the first edition with the same philosophy but clarifies the concept with recent research and thoughts of practitioners, added examples, added more industry methods, security issues and a discussion on regulatory issues.

The primary objective of this book is to provide a useful tool that can be used by any industrial company that handles hazardous chemicals to understand inherent safety concepts. Secondly, the book provides some tools and guidance on approaches to implement inherent safety.

---

[2] Center for Chemical Process Safety (CCPS) (1996). *Inherently Safer Chemical Processes - A Lifecycle Approach.* New York: American Institute of Chemical Engineers.

The book is intended for chemical site managers, process safety managers, engineers, chemists, regulators, engineering educators, and others responsible for chemical safety and interested in the application of inherent safety to process safety management.

The book covers the history, research, and basic concepts of inherent safety. In particular it includes guidance on how to conduct inherent safety studies and how to incorporate inherent safety into an organization's process safety management processes. The method described in this book may be widely applicable to inherent safety as it relates to safety, environment, and security issues.

Achievements of the Updated Concept Book

Besides a comprehensive update of the topic, the concept book improved the state of the art in inherent safety by making contributions in several areas:

1. the new concept of orders of Inherent Safety is introduced. This was intended to ease understanding of the various degrees of inherently safer actions that may be taken while considering any process risk, the recommended sequence of this consideration, and how inherent safety relates to layers of protection analysis. This helps define the type of inherently safer application, which will be constructive for discussions with practitioners, managers, and regulators. Figure 1 illustrates these concepts.
2. the appendices include illustrations of applying inherent safety across the entire life cycle
3. a new and more complete inherently safer checklist is presented.
4. practical methods of applying inherently safer strategies to a process and analyzing hazards and opportunities for risk reduction are included.

Organization of the Book

The book is written with the key principles for inherent safety in the body of the book, and tools for implementing the approach and worked examples and checklists in the appendices. The key chapters of the book are:

1. Introduction
2. The Concept of Inherent Safety
3. The Role of Inherently Safer Concepts in Process Risk Management
4. Inherently Safer Strategies
5. Life Cycle Stages
6. Human Factors
7. Inherent Safety and Security
8. Implementing Inherently Safer Design
9. Inherently Safer Design Conflicts
10. Inherently Safer Design Regulatory Initiatives

11. Worked Examples and Case Studies

12. Future Initiatives

Appendix: A Sample Inherently Safer Process Checklist

Chapter 2 introduces the topic of inherent safety. The key terms and the philosophy behind inherent safety are also described.

Inherently safer concepts will enhance overall risk management programs, whether directed toward reducing frequency or consequences of potential accidents. Ways in which inherent safety can be applied can be categorized into 'strategies'. These strategies -- minimize, substitute, moderate, and simplify -- are discussed in detail in this book in Chapters 3 and 4.

The process industry has recognized that a process goes through various stages of evolution. In this book, these stages are called life cycle stages as shown by Figure 1.2. The life cycle of a process begins with discovery at the research stage. Then a process progresses through stages of process development, design and construction, operations, maintenance, and modification. At the end, involvement with the process ends with decommissioning.

Exploring inherently safer alternatives may require more resources during the early stages of development than is otherwise the case. However, the resulting understanding will, in many cases, minimize or eliminate the need for appended safety mitigation devices and the costs of maintaining them, as well as reduce the possibility of incidents. Inherently safer considerations may reduce the life cycle cost of the process. In general, the economic benefits to be derived from inherently safer thinking and in some cases the feasibility of inherent safety, will increase by application early in the process. However, it is never too late to use inherently safer concepts for existing facilities, as it is likely that some positive risk reduction can be achieved even after the facility is operating.

Inherently safer is a way of thinking and to successfully implement it inherent safety has to be continually employed wherever possible. Improved understanding of the process may result in a better process and higher quality products. Processes should be reviewed for hazards and risks periodically. Chapter 5 discusses review methods to do this.

Human factors are an extremely important part of inherently safer concepts. Processes should be designed to reduce the opportunities for human error. Chapter 6 of this book presents a discussion of human factors as related to inherently safer design.

Chapter 7 discusses the role of inherent safety in chemical process security – a recent topic of interest and controversial issue for potential regulations.

Chapter 8 discusses available methods for implementing inherently safer strategies. These can either be independent, special studies done periodically or before a major project or change is undertaken or integral to day-to-day process risk management strategies and opportunistically applied.

Chapter 9, Inherently Safer Design Conflicts, describes the conflicts that often develop between various attributes of safety, operability, cost, and other risk parameters and the ways to understand and make decisions in light of those constraints.

With the advent of some state and local regulations and proposed regulations that require inherent safety consideration or implementation, and proposed Federal regulations for inherently safer design, Chapter 10 was written to help guide regulators and industry through the various considerations and challenges of IS.

Chapter 11 contains examples of IS study methods and some case studies to show the step-wise process that can be followed for an IS evaluation. It also gives practical examples of successful implementation.

Chapter 12 describes potential future IS initiatives including needs, research, expected practice issues, and regulatory issues. There is work to be done to improve the tools available for the application of inherently safer concepts.

The Concept of Inherent Safety

The modern approach to chemical process safety is to apply risk management systems theory. This includes recognition of the hazards posed by the process and a continual effort to analyze the risks and to reduce or control them to the lowest levels practical while considering the balance of other objectives of the business. A hazard is classically defined as a 'situation with the potential for harm'. Risk is defined as 'the likelihood that a defined consequence will occur'.

Process risk management is the term given to the collective efforts to manage process risks through a wide variety of strategies, techniques, procedures, policies, and systems. Chemical process hazards are defined to come from two sources:

- hazards that are characteristic of the materials and chemistry used, and
- hazards that are characteristic of the process variables -- the way the chemistry works in the process.

In general, the strategy for reducing risk, whether directed toward reducing the frequency or the consequences of potential accidents, can be classified into four categories. These categories are:

- *Inherent* - Eliminating the hazard by using materials and process conditions which are non-hazardous; e. g., substituting water for a flammable solvent.
- *Passive* - Minimizing the hazard by process and equipment design features which reduce either the frequency or consequence of the hazard without the active functioning of any device; e.g., providing a diked wall around a storage tank of flammable liquids..
- *Active* - Using controls, safety interlocks, and emergency shutdown systems to detect and correct process deviations; e.g. a pump which is shut-off by a high level switch in the downstream tank when the tank is 90% full. These systems are commonly referred to as engineering controls although human intervention is also an active layer.
- *Procedural* - Using policies, operating procedures, training, administrative checks, emergency response, and other management approaches to prevent incidents, or to minimize the effects of an incident; e.g. hot work procedures and permits. These approaches are commonly referred to as administrative controls.

Inherent Safety Defined

What do we mean when we speak of an "inherently safer" chemical process? "Inherent" has been defined as "existing in something as a permanent and inseparable element, quality, or attribute" (American College Dictionary, 1967). A chemical manufacturing process is inherently safer if it reduces or eliminates the hazards associated with materials and operations used in the process, and this reduction or elimination is permanent and inseparable. To appreciate this definition fully, it is essential to understand the precise meaning of the word "hazard." A hazard is defined as a physical or chemical characteristic that has the potential for causing harm to people, the environment, or property (adapted from CCPS, 1992). The key to this definition is that the hazard is intrinsic to the material, or to its conditions of storage or use. Some specific examples of hazards include:

- Chlorine is toxic by inhalation.
- Sulfuric acid is corrosive to the skin.
- Ethylene is flammable.
- Steam confined in a drum at 600 psig contains a significant amount of potential energy (PV and Thermal energy).
- Acrylic acid monomer can polymerize releasing large amounts of heat.

These hazards cannot be changed -- they are basic properties of the materials and the conditions of usage. The inherently safer approach is to reduce the consequences or likelihood of the hazard or by completely eliminating the hazardous agent.

For these reasons, the inherently safer approach should be an essential aspect of any process safety program. If the hazards can be eliminated or reduced, the extensive layers of protection to control those hazards will not be required or may be less robust.

Inherently Safer Approach

The essential issue with the concept of inherent safety is that the focus should be on reducing or eliminating hazards by changing the materials, chemistry, and process variables such that the reduced hazard is characteristic of the new conditions. This compares with adding layers of safety to a process to reduce the risk but not reducing the nature of the hazard directly.

The process with reduced hazards is described as *inherently safer*, rather than inherently safe, as it is a move in the direction of reducing the risk of realizing the consequences of concern. This terminology recognizes there is no chemical process that is without risk, but all chemical processes can be made safer by applying inherently safer concepts. This book occasionally uses the term "inherent safety"; this does not mean absolute safety. In any case the final goal is acceptable risk, where inherent safety could be one effective strategy to achieve that goal. Inherent safety is not the only process risk management strategy available and may not be the most effective. A system of strategies is applied to reduce risks to the lowest levels practicable.

The steps of analyzing, reducing, and managing risk should ideally be done in a hierarchical manner as shown in Figure 1. If feasible to implement inherently safer approaches alone to meet project risk goals, this may avoid costs associated with time, capital, and expense for the layers of protection that would otherwise be required.

Layers of Protection

The other strategies of Passive, Active, and Procedural are considered 'layers of protection' as they involve the addition of safety devices or work processes to reduce risks. Passive safety devices do not perform any fundamental operation, but are designed to be available when a process upset occurs. Procedural safety measures, or administrative controls, utilize safe work practices and procedures to reduce risk. On the other hand, inherent safety uses the properties of a material or process to eliminate or reduce the hazard. The fundamental difference between inherent safety and the other three categories is that inherent safety seeks to remove the hazard at the source as opposed to accepting the hazard and attempting to mitigate the effects.

'Layers of protection' is a concept whereby several different devices, systems, or actions are provided to reduce the likelihood and severity of an undesirable event. This concept is based on the premise that for an undesired event to occur, a number of protective features and countermeasures must fail, assuming that appropriate layers (or barriers) have been designed into the process or site. In order to be considered adequate there must be an adequate number of Independent Protection Layers (IPLs), which include any device, system, or action that is capable of preventing a scenario from proceeding to the undesired consequence regardless of the initiating event or the action of any other protection layer associated with the scenario.

These layers of protection may include operator supervision, control systems, alarms, interlocks, physical protection devices, and emergency response systems. Commonly referred to as 'barriers', they may include far more than physical barriers. This approach can be highly effective, and its application has resulted in significant improvement in the safety record of the chemical industry.

The approach of imposing barriers between a hazard and potentially impacted people, property, and environment has significant disadvantages:

- The hazard remains, and some combination of failures of the layers of protection may result in an incident thereby allowing the hazard to be fully realized. Every layer has a certain likelihood of failure due to mechanical means or failures of management systems, such as failure to maintain or to keep administrative controls active. The outcome of the event may be limited to whatever passive or inherent layers have been applied. If the overall risk was justified to be low in consideration of those layers, there could be substantial residual consequences.
- Because the hazard is still present, there is always a danger that its potential impacts could be realized by some unanticipated route or mechanism. Nature may be more creative in inventing ways by which a hazardous event can occur than experts are in identifying them. Accidents can occur by mechanisms that were unanticipated or poorly understood.

The layers of protection can be expensive to build and maintain throughout the life of the process. Factors include initial capital expense, operating costs, safety training cost, maintenance cost, and diversion of scarce and valuable technical resources into maintenance and operation of the layers of protection.

Levels of Inherent Safety

The steps of managing risk should ideally be done in a hierarchical manner and iteratively as shown in Figure 1 (adapted from Amyotte, et. al, 2006). The process risk management approach illustrated begins with the definition of goals for managing risk.

Without a clear definition the need for further investment in safety is unclear short of any regulatory requirements. This is important in determining how 'inherently safe is safe enough'.

There can be much discussion about whether or not a particular safety feature in a chemical process is "inherent." Such discussions may arise in part because different people have different perspectives on risk or are viewing the process at different levels of resolution, ranging from a global view of the entire process to a very detailed view of specific features of the process. Also, they may be referring to addressing the hazards of the process vs. addressing other aspects of managing the risk of the process hazards.

In the strictest sense (***or the 1st order view of inherent safety***), one could argue that the definition of inherently safer applies only to the elimination of a hazard. Inherently safer strategies may absolutely eliminate a hazard (and hopefully not introduce another hazard of concern as a result).

Alternatively, inherently safer approaches could instead treat the hazard by making it less intense or likely to occur (***or the 2nd order view of inherent safety***). This is clearly in line with inherent safety philosophy but may not be as powerful as a 1st order change. In the 2nd order of IS the hazard is only reduced through the application of IS principles. It could be that 2nd order inherently safer design options result in a considerable reduction of hazard and therefore the risk is adequately addressed.

In the broadest sense the overall hazard is not eliminated or reduced by way of inherently safer strategies but instead sublevel hazards are minimized and the likelihood of the event occurring is reduced by adding layers of protection. The strength and reliability of a layer of protection can vary, i.e., more robust layers. Any layer could be chosen to be more 'robust' than another, which could mean it is more reliable, effective, or simpler, or other positive safety attributes are achieved by comparison, but the fundamental hazard may still exist. This is the difference between the inherently safer design concepts being applied to the hazard and layers of protection being applied to reduce the overall risk.

Inherent Safety Concepts - Summary

Inherently safer design is a fundamentally different way of thinking about the design of chemical processes and plants. It focuses on the elimination or reduction of the hazards, rather than on management and control. This approach should result in safer and more robust processes, and it is likely that these inherently safer processes will also be more economical in the long run (Kletz, 1984, 1991b). It must be recognized, however, that an inherently safer plant is not necessarily the most efficient or economical, as the cost of changing an existing design to a more inherently safer technology may be unjustified or difficult to justify from an investment standpoint. For this reason, the options must be holistically weighed and the total lifecycle costs and risks be analyzed for completeness.

Eliminating or reducing the hazard through the application of ISD is the first process risk management strategy. Other strategies involve adding safety layers to the hazard to reduce likelihood or consequences. Thus, layers of protection can be classified into three categories, generally listed in decreasing order of reliability and robustness: passive, active, and procedural.

Figure 1
Inherently Safer Approach to Analyzing and Managing Process Risks

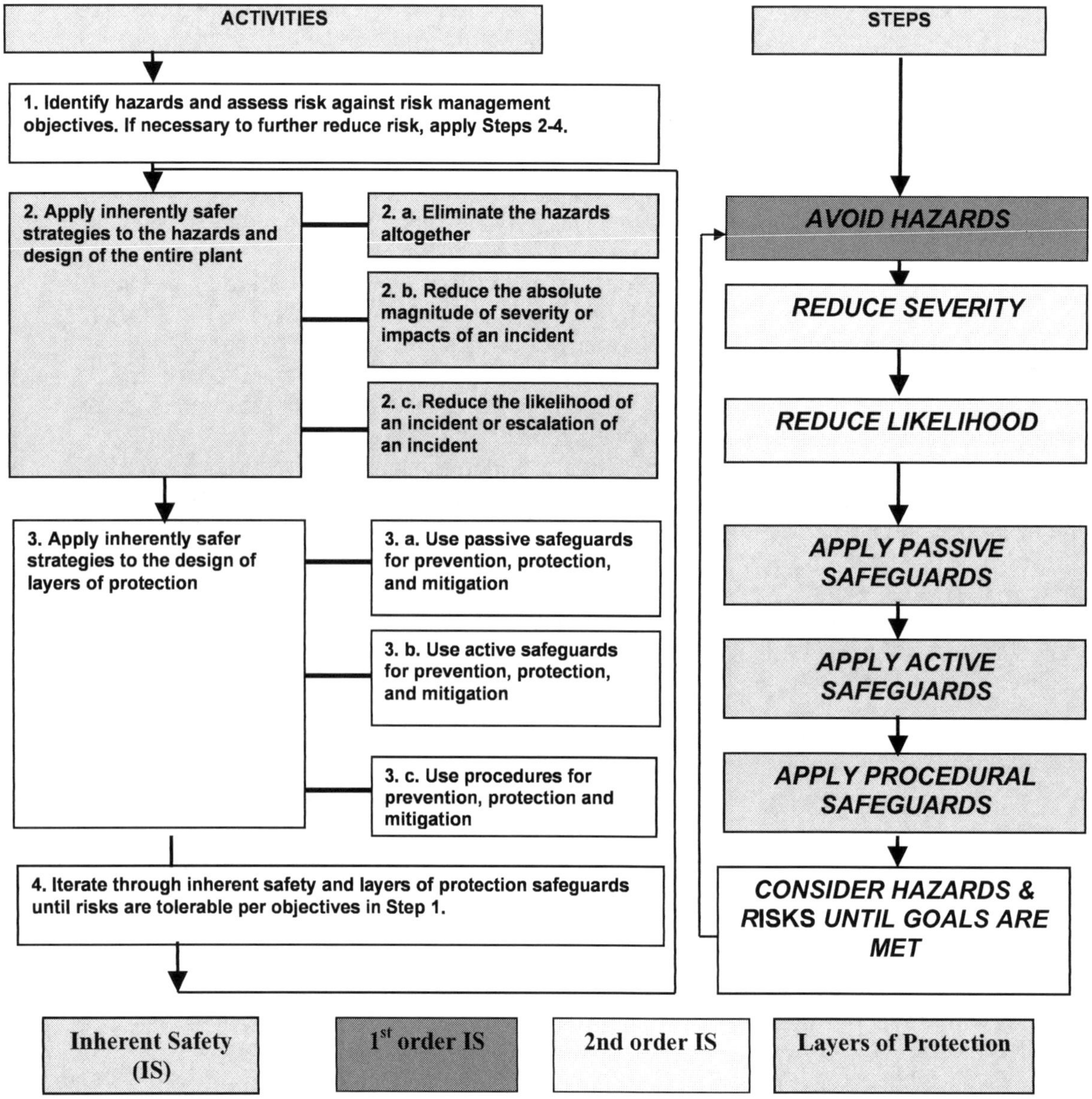

Legal Consideration of Risk Tolerance

This presentation will review the legal issues associated with risk tolerance. It will focus on "acceptable risk" within the context of OSHA and examine both qualitative and quantitative risk assessment. Lastly, it will discuss two case studies of significant incidents in which risk tolerance was a significant factor in the defense of the OSHA citations.

Mark S. Dreux
Arent Fox PLLC
1050 Connecticut Avenue, N.W.
Washington, D.C. 20036
dreux.mark@arentfox.com
Phone: (202) 857-6405
Fax: (202) 857-6395

AUTHOR INDEX

American Institute of Chemical Engineers
3 Park Avenue
New York, NY 10016-5991

ISBN: 978-1-60423-310-0